# Nanomaterials for Surface-Enhanced Raman Spectroscopy and Application in Trace Detection

# Nanomaterials for Surface-Enhanced Raman Spectroscopy and Application in Trace Detection

Special Issue Editors

**Xiaonan Lu**

**Yaxi Hu**

MDPI • Basel • Beijing • Wuhan • Barcelona • Belgrade • Manchester • Tokyo • Cluj • Tianjin

*Special Issue Editors*
Xiaonan Lu
University of British Columbia
Canada

Yaxi Hu
University of British Columbia
Canada

*Editorial Office*
MDPI
St. Alban-Anlage 66
4052 Basel, Switzerland

This is a reprint of articles from the Special Issue published online in the open access journal *Nanomaterials* (ISSN 2079-4991) (available at: https://www.mdpi.com/journal/nanomaterials/special_issues/raman_detect).

For citation purposes, cite each article independently as indicated on the article page online and as indicated below:

LastName, A.A.; LastName, B.B.; LastName, C.C. Article Title. *Journal Name* **Year**, *Article Number*, Page Range.

**ISBN 978-3-03936-272-1 (Hbk)**
**ISBN 978-3-03936-273-8 (PDF)**

# Contents

# About the Special Issue Editors

**Xiaonan Lu** (Associate Professor) received his B.Sc. (Food Science) from Ocean University of China in 2007 and Ph.D. (Food Science) from Washington State University in 2011. He was a postdoctoral fellow in the College of Veterinary Medicine at Washington State University before he joined in UBC in January 2013, as the director of the UBC Food Safety Engineering Laboratory. He is the recipient of UBC Peter Wall Scholar (2017), the recipient of Young Scientist Excellence Award from International Union of Food Science and Technology (2015), the recipient of Young Scientist Travel Award from Agricultural & Food Chemistry Division of the American Chemical Society (2014), and currently the editorial board member of *Applied and Environmental Microbiology, Journal of Food Science, Food and Agricultural Immunology*, etc. He is the author of over 100 scientific papers as well as one book (Sensing Techniques for Food Safety and Quality Control, Royal Society of Chemistry) and several book chapters.

**Yaxi Hu** (Postdoctoral Fellow) received her B.Sc. in Biological Science from China Agricultural University (2013) and completed her Ph.D. in Food Science (2018) at The University of British Columbia. Caring about the food system's integrity, during her graduate studies, Dr. Hu focused on the development and fabrication of novel sensor- and instrument-based analytical techniques to detect food safety, food quality and food fraud related issues. The cutting edge technologies involved in her studies cover a broad range of techniques including paper- and polymer-based microfluidic "lab-on-a-chip" devices, nanomaterials, molecularly imprinted polymers, immunological assays, surface-enhanced Raman spectroscopy, and etc. Using these tools, Dr. Hu has published 20 peer-reviewed journal articles, with two additional manuscripts submitted, and three more manuscripts under preparation. Dr. Hu's research has been recognized by the Mitacs Outstanding Innovation-Ph.D. 2016, won her the first place in the Withycombe-Charalambous Graduate Student Symposium (2018, American Chemical Society—Agricultural and Food Division), and featured in many media, such as CBC, CTV, Global News, and others.

# Preface to "Nanomaterials for Surface-Enhanced Raman Spectroscopy and Application in Trace Detection"

Rapid and sensitive detection and quantification of trace-level chemicals is always of great interests, because the ability of analyzing the low concentration compounds in complex sample matrices, such as environmental, biological and food samples, enables improved understanding of these samples and how the samples representing different systems may interact with each other. With the advances in nanomaterials and nanofabrication, surface-enhanced Raman spectroscopy (SERS) has been extensively developed and applied in the trace detection of various analytes in either a simple or a complicated sample matrix. This includes, but is not limited to, the detection of antibiotic residues in animal-producing meat products, detection of pathogenic bacteria in human body fluid, and detection of heavy metal contamination of water. This book, consisting two review articles and five research articles, covers the most recent progress and advancement in the development and application of various nanomaterials in SERS trace detection. In this book, a broad range of topics is covered, from the synthesis of novel nanomaterials that can provide improved reproducibility of SERS signals to the development of new application protocols that can facilitate the reliable detection of trace amounts of analytes without interfered by the sample matrices significantly. This book is a useful source for both new and advanced researchers in the field of SERS and its application. Hope you enjoy the reading!

**Xiaonan Lu, Yaxi Hu**
*Special Issue Editors*

*nanomaterials*

MDPI

*Article*

# Large-Scale Fabrication of Ultrasensitive and Uniform Surface-Enhanced Raman Scattering Substrates for the Trace Detection of Pesticides

**Jia Zhu [1], Guanzhou Lin [1,2], Meizhang Wu [3], Zhuojie Chen [1], Peimin Lu [2] and Wengang Wu [1,*]**

1 National Key Laboratory of Science and Technology on Micro/Nano Fabrication, Institute of Microelectronics, Peking University, Beijing 100871, China; zhujia@pku.edu.cn (J.Z.); tk968810@163.com (G.L.); geoforchinky@163.com (Z.C.)

2 College of Physics and Information Engineering, Fuzhou University, Fuzhou 350116, China; lpm@fzu.edu.cn

3 The Affiliated High School of Peking University, Beijing 100080, China; wumeizhang2001@126.com

* Correspondence: wuwg@pku.edu.cn; Tel.: +86-10-6276-7553

Received: 8 June 2018; Accepted: 9 July 2018; Published: 12 July 2018

**Abstract:** Technology transfer from laboratory into practical application needs to meet the demands of economic viability and operational simplicity. This paper reports a simple and convenient strategy to fabricate large-scale and ultrasensitive surface-enhanced Raman scattering (SERS) substrates. In this strategy, no toxic chemicals or sophisticated instruments are required to fabricate the SERS substrates. On one hand, Ag nanoparticles (NPs) with relatively uniform size were synthesized using the modified Tollens method, which employs an ultra-low concentration of $Ag^+$ and excessive amounts of glucose as a reducing agent. On the other hand, when a drop of the colloidal Ag NPs dries on a horizontal solid surface, the droplet becomes ropy, turns into a layered structure under gravity, and hardens. During evaporation, capillary flow was burdened by viscidity resistance from the ropy glucose solution. Thus, the coffee-ring effect is eliminated, leading to a uniform deposition of Ag NPs. With this method, flat Ag NPs-based SERS active films were formed in array-well plates defined by hole-shaped polydimethylsiloxane (PDMS) structures bonded on glass substrates, which were made for convenient detection. The strong SERS activity of these substrates allowed us to reach detection limits down to $10^{-14}$ M of Rhodamine 6 G and $10^{-10}$ M of thiram (pesticide).

**Keywords:** SERS; Ag NPs; coffee ring; pesticide detection

## 1. Introduction

After several decades of development since it was discovered on electrochemically roughened silver in 1973 [1,2], surface-enhanced Raman scattering (SERS) has become a powerful analytical tool for applications of chemical and biological molecule detection, environmental monitoring, and food safety [3–8]. SERS is able to identify molecules through vibrational fingerprint signals and can even detect single molecules [9,10]. It is well accepted that a Raman signal can be enormously enhanced by noble metal nanostructures with sub–10 nm gaps between them, which we call 'hot spots' [9] Over the past decades, significant efforts in the areas of electron beam lithography [11], colloidal lithography [12], chemical synthesis [13–15], and self-assembly [16–18] have been made to develop highly active SERS substrates. All of these efforts are been focused on sufficiently high electromagnetic field enhancement, good SERS signal stability, and convenience in fabrication and manipulation. However, the above-mentioned requirements are hardly being met simultaneously. Electron beam lithography, nanoimprint lithography, and colloidal lithography can fabricate highly-uniform Ag or Au nanostructures leading to stable and reproducible SERS signals, but these methods are generally expensive and time consuming for large-scale fabrication. Ag or Au nanoparticles (NPs) formed by chemical synthesis is a very popular approach to SERS substrate preparation, and tremendous Raman enhancement could be achieved effortlessly by rich 'hot spots'. However, the downside of this kind of method is that the stability cannot be guaranteed because of the uneven distribution of 'hot spots'. For example, the NPs often aggregate in solution, which is not conducive to long-term preservation [19]. With regard to the SERS measurement, a general and simple method is mixing analytes with NP solutions, and then measuring the mixtures directly [20]. However, this method is incapable of trace detection. An improved method is to dry the mixtures. It is true that the NPs will be closely packed after drying [21–24], but this gives rise to a challenge of terrible aggregation caused by the coffee ring effect, also resulting in signal instability. Although, a coating method has been proposed to realize uniform and high-density Ag NPs distribution in drying process [25], this method suffers from oxidation of Ag NPs as time goes on. Therefore, it is still a great challenge to fabricate large-scale SERS substrates with uniform and high-density hot spots via simple and low-cost strategies.

The coffee-ring is a pattern left by a puddle of a particle-laden liquid after evaporation, which is almost familiar to everyone [26]. It is difficult to eliminate this ubiquitous effect from many applications, including the printing, assembly, and distribution of nano/molecular materials [27,28]. Closely packing Ag or Au NPs is the easiest way to obtain SERS substrates that might have a substantial enhancement of detection signals. The coffee-ring effect will make the Ag or Au NPs form as a ring, so that the distribution of the 'hot spots' is nonuniform and uncontrollable [22,23,29].

Herein, we present a convenient and inexpensive strategy to fabricate large-scale SERS substrates with stable and ultrasensitive performance. It involves a green chemistry synthesis method of Ag NPs and a facile approach of dropping the Ag NPs/glucose solution to form a flat film array for SERS detection. Viscous forces from the ropy glucose suppresses the coffee-ring effect, and thus leads to a uniform and compact deposition, but not aggregation of Ag NPs. Due to the wettability of the Ag NPs/glucose film, uniform distribution of analytes is also realized. These make the SERS signal more consistent and sensitive. In this strategy, no toxic chemicals or sophisticated instruments are required to fabricate the SERS substrate. In addition, thanks to the protection of glucose, oxidation of the Ag NPs is avoided, which results in their long-term storage (at least 6 months). Finally, we demonstrate the application of such SERS substrates for detection of R6G (Rhodamine 6G) and thiram (pesticide) down to $10^{-14}$ M and $10^{-10}$ M, respectively.

## 2. Materials and Methods

### 2.1. Materials

Silver nitrate (99.9%), R6G ($C_{28}H_{31}N_2O_3Cl$, 99%) and thiram ($C_6H_{12}N_2S_2$, 99.9%) were purchased from Sigma-Aldrich (Darmstdt, Garmany), Ammonia (25% $w/w$ aqueous solution) and D-glucose were supplied by Beijing Chemical Works (Beijing, China), and SYLGARD 184 Silicone Elastomer Base and SYLGARD 184 Silicone Elastomer Curing Agent were purchased from Dow Corning Corporation (Midland, MI, USA). All the reagents used in this work were of analytical grade. Deionized water (Milli-Q purification system, Millipore Co., Bedford, MA, USA) was used for all experiments.

### 2.2. Characterizations

UV-visible spectra were recorded with a 1 cm path length quartz cell using an Agilent Cary 8454 spectroscopy system (Agilent Technologies Inc., Santa Clara, CA, USA). Polydimethylsiloxane (PDMS) surfaces were treated with a BD-20AC laboratory corona treater (Electro-Technic Products Inc., Chicago, IL, USA). Scanning transmission electron microscope (STEM) measurements were conducted on Tecnai G2 F20 (FEI, Hillsboro, OR, USA). The Raman spectra were obtained using a Renishaw inVia Reflex Raman Microscope and Spectrometer (inVia Reflex, Gloucestershire, UK) equipped with a 633 nm laser and 50× objective. The integration time of all spectra acquisition for each measurement was set to be 10 s. The laser power was 1.7 mW. Four spots on the same SERS substrate were examined, and the spectra were averaged for final analysis.

### 2.3. Preparation of Ag NPs

The Ag NPs were prepared according to our previous reports [30]. Briefly, approximately 1 mL of ammonium hydroxide was added drop by drop into a fast stirring silver nitrate solution (85 mg in 10 mL water). A dark-brownish precipitate was formed and then dissolved as the amount of ammonium hydroxide is increased. Subsequently, a small amount of as-prepared Tollens solution (40 µL) was added into a fresh solution (40 mL) of D-glucose with a concentration of 0.1 M to 0.5 M. After four hours at room temperature, $Ag^+$ ions were reduced by glucose in the presence of ammonia, and the color of the glucose solution turned to yellow, indicating the formation of Ag NPs [31].

### 2.4. Fabrication of SERS Substrates

Firstly, the thin PDMS films with a thickness of 0.5 mm were prepared by mixing a Slygard 184 elastomer with a curing agent in a 10:1 ratio at 70 °C for two hours. Then the PDMS films were punched to form an array of holes with a diameter of 5 mm. Subsequently, the punched PDMS films were treated by $O_2$ plasma using a BD-20AC laboratory corona treater (Electro-Technic Products Inc., Chicago, IL, USA) for 30 s. Then, the treated PDMS surface was bonded to a glass slide under 70 °C for 2 h. Finally, 100 µL of prepared colloidal Ag NPs were dropped into each hole of the PDMS films. After the droplets were evaporated, the pie-shaped substrates of Ag NPs were formed, and then stored at room temperature for SERS detecting. The fabricated SERS film array and the SERS detection process is shown in the Figure 1.

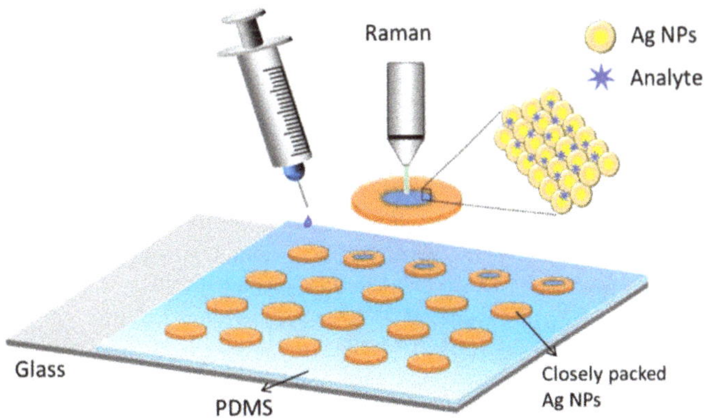


**Figure 1.** Schematics of the SERS detection process. The Ag NPs are closely packed and surrounded by massive solid-state glucose. Moreover, analytes can be distributed uniformly in the SERS substrates by infiltration and capillarity.

## 3. Results and Discussion

### 3.1. Characterization of Ag NPs

UV-vis absorption spectra of the colloidal Ag NPs synthesized in 0.5 M glucose is given in Figure 2a. The absorption peak was around 415 nm, exhibiting a sharp plasmon absorption maximum. It indicated that the Ag NPs were monodispersed and relatively uniform in a stable colloidal solution [30,31]. A TEM image of Ag NPs is given in Figure 2b. The size of the Ag NPs was 35 ± 3 nm (the UV-vis absorption spectra and TEM images of the Ag NPs synthesized in 0.1–0.5 M glucose are given in Figures S1 and S2). Since the amount of glucose is excessive, the $Ag^+$ ions have been completely reduced. Moreover, it was a green and highly-efficient synthesis method that does not need a heating condition or any relatively toxic organic surfactants.

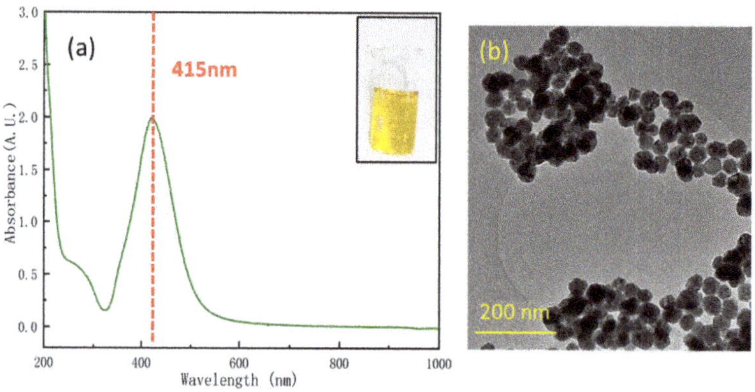


**Figure 2.** (**a**) UV-vis absorption spectra of the Ag NPs, and the inset shows a photograph of the Ag NP colloid, and (**b**) TEM image of the Ag NPs, and the size of Ag NPs were 35 ± 3 nm.

### 3.2. Suppression of the Coffee-Ring Effect/Pie-Shaped SERS Substrates

When a drop of liquid dries on a solid surface, the liquid evaporates first from the edge and is replenished by liquid from the interior, which results in capillary flow in the drop during its drying process. The suspended particles are driven to the edge by the capillary flow, and then left highly concentrated along the original drop edge, finally depositing in a ring-like pattern after evaporation [28]. Many attempts to suppress and ameliorate the coffee ring effect have thus far focused on manipulating the capillary flow [32–35]. In our strategy, the solution became viscous owing to the existence of massive glucose during the evaporation of the as-prepared colloidal Ag NPs, and the capillary flow is therefore burdened by viscidity resistance, which prevented the suspended particles from reaching the drop edge and ensured their uniform deposition. Figure 3a shows the evaporation process of a drop of as-prepared colloidal Ag NPs dropped onto a clear glass directly. As the water evaporates, the concentration of the glucose is increased, which resulted in an increasing viscosity of the solution [36]. Finally, the droplet had turned into a rigid film and the coffee-ring effect was almost eliminated after evaporation. As the droplet turned to a rigid film, the volume shrunk dramatically, therefore the gap between the Ag NPs would decrease and even turn into an aggregate, which exhibits higher SERS activity [37]. In order to make the rigid film well-shaped and standardize the fabricating process, the colloidal Ag NPs were dropped into an array of holes, which were defined by PDMS structures bonded on the glass substrates (Figure 3b). Figure S3 shows the extinction spectrum of the SERS substrate with 0.2 M glucose. Thanks to the protection of glucose, oxidation of the Ag NPs was also prevented, which resulted in superior stability of the SERS substrates. From the test, the stability and sensitivity of SERS signals could remain for at least 6 months (Figures S4 and S5). The characteristic peaks of R6G were still very distinct after 6 months, and the intensity of the peaks was comparable to the intensity when the substrate was deposited.

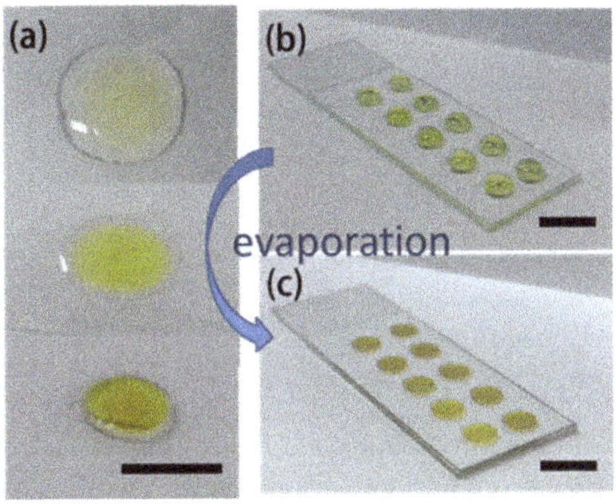


**Figure 3.** (**a**) The evaporation process of a drop of the solution, and (**b**) a photograph of colloidal Ag NPs dropped into an array of holes. The holes were 5 mm in diameter, 1 mm in depth, and 10 mm apart for standardized fabrication and convenient measurement of the SERS films in array. (**c**) The fabricated pie-shaped SERS substrates after evaporation. The scale bar is 1 cm.

### *3.3. Optimal Concentration of Glucose*

According to the enhancement mechanism of SERS signals, more sensitive SERS signal could be realized when analytes are close enough to the Ag NPs surface. In our strategy, the glucose protected the Ag NPs from oxidation while preventing the touch between the analyte and the Ag NPs to some extent. Therefore, the optimal concentration of glucose is of great necessity. In experiments, colloidal Ag NPs with different concentrations of glucose, ranging from 0.1 M to 0.5 M, were used to fabricate the SERS substrates. R6G ($10^{-8}$ M) was employed as a SERS probe to evaluate the performance of these SERS substrates. As shown in Figure 4, the SERS signal intensity of these five substrates was in the order: 0.2 M > 0.1 M > 0.3 M > 0.4 M > 0.5 M, suggesting the substrate fabricated with a 0.2 M concentration of glucose showed a stronger signal than the others. When 1 μL of R6G solution was dropped onto the SERS substrates, the water would dissolve the glucose that covered the Ag NPs, and then the Ag NPs were exposed to analytes. Hence, a lower concentration of glucose was more likely to supply bare Ag NPs, resulting in a higher Raman enhancement. As shown in Figure S6, there was a risk of oxidation of Ag NPs with a much lower concentration of glucose (such as 0.1 M). Accordingly, we choose a 0.2 M Ag NPs/glucose solution to fabricate the SERS substrates for later detection.

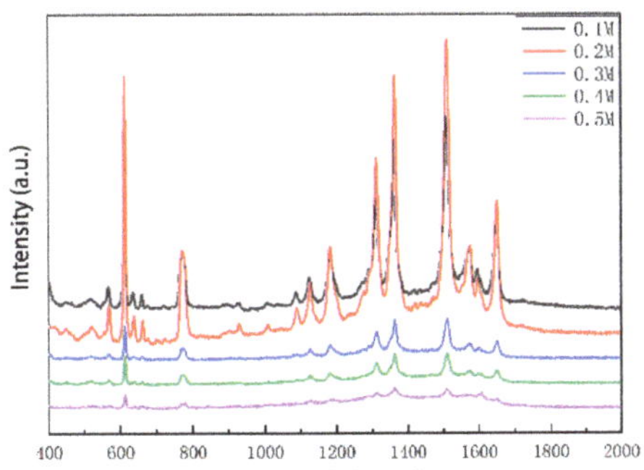


**Figure 4.** SERS spectra of $10^{-8}$ M R6G as different concentrations of glucose are used to fabricate the SERS substrates.

### 3.4. Sensitivity and Reproducibility of SERS Substrates

Sensitivity and reproducibility are major concerns for any SERS substrates. Hence, we employed different concentrations of R6G to investigate the performance of the SERS substrates. As shown in Figure 5a, the strong peaks at 612 $cm^{-1}$, 774 $cm^{-1}$, 1127 $cm^{-1}$, 1185 $cm^{-1}$, 1310 $cm^{-1}$, 1360 $cm^{-1}$, 1509 $cm^{-1}$, 1573 $cm^{-1}$, and 1650 $cm^{-1}$ were in good agreement with previous reports on pure R6G [38–40]. These SERS spectral feature peaks can still be clearly identified even when the concentration was down to $10^{-14}$ M, corresponding to 10 zeptomoles of R6G molecules in a 1 μL sample volume, which is far below previous works [38–41]. The reproducibility of the SERS substrate was further investigated by taking SERS spectra of R6G at the concentration of $10^{-10}$ M from 20 random locations on a single pie-shaped substrate. The average relative standard deviation (RSD) of the intensities (at 1509 $cm^{-1}$) was 6.8% (Figure 5b and Figure S7) which is lower than previously reported [40,42], indicating that the substrates possess good signal uniformity.

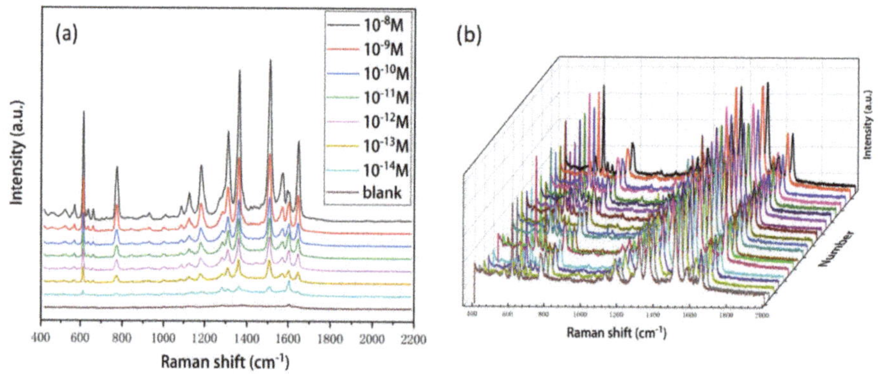


**Figure 5.** (**a**) SERS spectra of R6G at a concentration of $10^{-8}$ M to$10^{-14}$ M. 1 μL of the R6G solution was dropped to the SERS substrates for detection. The blank is a control group which used deionized water. (**b**) Raman spectra of $10^{-10}$ M R6G collected from 20 random spots. The RSD of the intensity maximum at the peak of 1509 $cm^{-1}$ was 6.8%.

### 3.5. Application for Thiram Detection

Thiram is a typical sulfur-containing pesticide molecule, which is widely used in agriculture. In this contribution, the SERS substrates were employed to detect thiram for future practical applications. The SERS spectra of thiram with a concentration varying from $10^{-6}$ M to $10^{-10}$ M are shown in Figure 6. The SERS signal intensity gradually increased with the increase of thiram concentration, pointing to the possibility of quantitative analyte determination. The main Raman bands include 563 $cm^{-1}$ attributed to υ(S-S), 1147 $cm^{-1}$ corresponding to $\rho(CH_3)$ and υ(C-N), and 1383 $cm^{-1}$ corresponding to $\delta_s(CH_3)$ and υ(C-N), and 1511 $cm^{-1}$ corresponding to υ(C-N), $\delta(CH_3)$, and $\rho(CH_3)$ [42,43]. We could identify the spectrum of the thiram even at the concentration of $10^{-10}$ M. The result suggested that the ability of SERS substrates to have a detection sensitivity of $10^{-10}$ M of thiram, which is far below the maximum residue limits of thiram for vegetables (5 mg/kg, equal to $\approx 2 \times 10^{-5}$ M). It is evident that such SERS substrates would have great potential for real world applications.

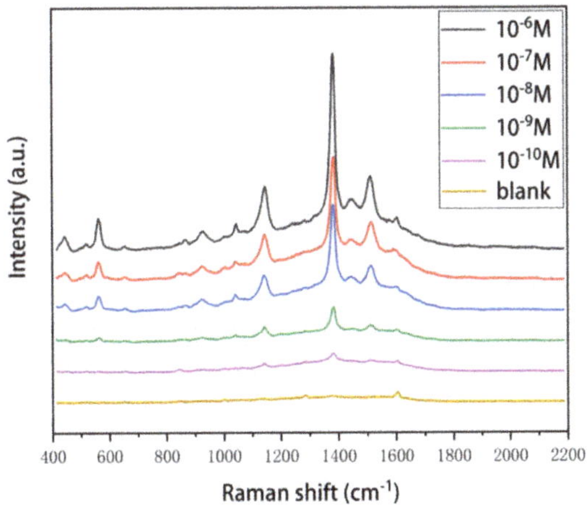


**Figure 6.** SERS spectra of thiram at a concentration range from $10^{-6}$ M to $10^{-10}$ M. The blank is a control group with DI water.

## 4. Conclusions

In summary, we have demonstrated a method for a simple and economically viable design of large-scale, highly efficient, ultrasensitive, uniform, and low-cost SERS substrates, especially emphasizing the suppression of the coffee-ring effect. In this method, the glucose acts to achieve this suppression. First, Ag NPs were synthesized by a modified Tollens method in which glucose serves as a reducing agent. Secondly, a Ag NPs/glucose solution was dropped into a hole-shaped PDMS structure. After evaporation, a flat and uniform SERS film array was formed by means of viscidity resistance in a ropy glucose solution, which prevented the suspended particles from reaching the drop edge and ensured a uniform deposition. Thirdly, owing to the protection of glucose, oxidation of the Ag NPs was also avoided, which resulted in long-term storage (at least 6 months) of the SERS substrates. Subsequently, the performance of the SERS substrates fabricated with different concentrations of glucose was investigated, and the result suggested that the optimal concentration of glucose was 0.2 M. Finally, we demonstrated the application of such SERS substrates for detection of R6G and thiram down to $10^{-14}$ M and $10^{-10}$ M, respectively. Thus, such a convenient fabrication method and superior performance of the obtained SERS substrates would provide an opportunity to bring the SERS technology closer to real-world applications.

**Supplementary Materials:** The following are available online at http://www.mdpi.com/2079-4991/8/7/520/s1, Figure S1: UV-vis absorption spectra of the colloidal Ag NPs synthesized in 0.1–0.5 M glucose; Figure S2: TEM images of the Ag NPs synthesized in 0.1–0.5 M glucose; Figure S3: SERS intensity of $10^{-10}$ M R6G collected at 1509 $cm^{-1}$ from 20 random spots of a pie-shaped SERS substrate; Figure S4: SERS spectra of R6G ($10^{-10}$ M) collected from the SERS substrates at different storage times ((a) 0 months, (b) 2 months, (c) 4 months, and (d) 6 months).

**Author Contributions:** Conceptualization, J.Z. and W.W.; Methodology, J.Z.; Software, G.L. and M.W.; Investigation, J.Z. and G.L.; Writing-Original Draft Preparation, J.Z.; Writing-Review & Editing, W.W. and P.L.; Visualization, Z.C.; Project Administration, W.W.; Funding Acquisition, W.W., All authors read and approved the final manuscript.

**Funding:** This research was funded by the National Basic Research Program of China (973 Program, Grant No. 2015CB352100), the National Natural Science Foundation of China (Grant No. 61474006), and the National Key Laboratory Foundation.

**Acknowledgments:** We thank the Electron Microscopy Laboratory of Peking University for the use of their equipment.

**Conflicts of Interest:** The authors declare no conflict of interest.

## References


1. Fleischmann, M.; Hendra, P.J.; McQuillan, A.J. Raman spectra of pyridine adsorbed at a silver electrode. *Chem. Phys. Lett.* **1974**, *26*, 163–166. [CrossRef]
2. Jeanmaire, D.L.; Van Duyne, R.P. Surface Raman spectroelectrochemistry: Part I. Heterocyclic, aromatic, and aliphatic amines adsorbed on the anodized silver electrode. *J. Electr. Chem. Interfacial Electrochem.* **1977**, *84*, 1–20. [CrossRef]
3. Tian, S.; Neumann, O.; McClain, M.J.; Yang, X.; Zhou, L.; Zhang, C.; Nordlander, P.; Halas, N.J. Aluminum nanocrystals: A sustainable substrate for quantitative SERS-based DNA detection. *Nano Lett.* **2017**, *17*, 5071–5077. [CrossRef] [PubMed]
4. Park, M.; Jung, H.; Jeong, Y.; Jeong, K.H. Plasmonic schirmer strip for human tear-based gouty arthritis diagnosis using surface-enhanced Raman scattering. *ACS Nano* **2017**, *11*, 438–443. [CrossRef] [PubMed]
5. Zhao, Z.; Huang, Y.; Fan, Y.; Lai, K. Rapid detection of flusilazole in pears with Au@Ag nanoparticles for surface-enhanced Raman scattering. *Nanomaterials* **2018**, *8*, 94. [CrossRef] [PubMed]
6. Hakonen, A.; Rindzevicius, T.; Schmidt, M.S.; Andersson, P.O.; Juhlin, L.; Svedendahl, M.; Boisen, A.; Kall, M. Detection of nerve gases using surface-enhanced Raman scattering substrates with high droplet adhesion. *Nanoscale* **2016**, *8*, 1305–1308. [CrossRef] [PubMed]
7. Gao, F.; Liu, L.; Cui, G.; Xu, L.; Wu, X.; Kuang, H.; Xu, C. Regioselective plasmonic nano-assemblies for bimodal sub-femtomolar dopamine detection. *Nanoscale* **2017**, *9*, 223–229. [CrossRef] [PubMed]
8. Wang, P.; Pang, S.; Chen, J.; McLandsborough, L.; Nugen, S.R.; Fan, M.; He, L. Label-free mapping of single bacterial cells using surface-enhanced Raman spectroscopy. *Analyst* **2016**, *141*, 1356–1362. [CrossRef] [PubMed]
9. Zhang, R.; Zhang, Y.; Dong, Z.C.; Jiang, S.; Zhang, C.; Chen, L.G.; Zhang, L.; Liao, Y.; Aizpurua, J.; Luo, Y.; et al. Chemical mapping of a single molecule by plasmon-enhanced Raman scattering. *Nature* **2013**, *498*, 82–86. [CrossRef] [PubMed]
10. Zrimsek, A.B.; Chiang, N.; Mattei, M.; Zaleski, S.; McAnally, M.O.; Chapman, C.T.; Henry, A.I.; Schatz, G.C.; Van Duyne, R.P. Single-molecule chemistry with surface- and tip-enhanced Raman spectroscopy. *Chem. Rev.* **2017**, *117*, 7583–7613. [CrossRef] [PubMed]
11. Gutierrez-Rivera, L.; Peters, R.F.; Dew, S.K.; Stepanova, M. Application of EBL fabricated nanostructured substrates for surface enhanced Raman spectroscopy detection of protein A in aqueous solution. *J. Vac. Sci. Technol. B* **2013**, *31*, 06F901. [CrossRef]
12. Zhu, C.; Meng, G.; Zheng, P.; Huang, Q.; Li, Z.; Hu, X.; Wang, X.; Huang, Z.; Li, F.; Wu, N. A hierarchically ordered array of silver-nanorod bundles for surface-enhanced Raman scattering detection of phenolic pollutants. *Adv. Mater.* **2016**, *28*, 4871–4876. [CrossRef] [PubMed]
13. Zhong, L.B.; Yin, J.; Zheng, Y.M.; Liu, Q.; Cheng, X.X.; Luo, F.H. Self-assembly of Au nanoparticles on PMMA template as flexible, transparent, and highly active SERS substrates. *Anal. Chem.* **2014**, *86*, 6262–6267. [CrossRef] [PubMed]
14. Polavarapu, L.; Mourdikoudis, S.; Pastoriza-Santos, I.; Pérez-Juste, J. Nanocrystal engineering of noble metals and metal chalcogenides: Controlling the morphology, composition and crystallinity. *Cryst. Eng. Commun.* **2015**, *17*, 3727–3762. [CrossRef]
15. Novara, C.; Dalla Marta, S.; Virga, A.; Lamberti, A.; Angelini, A.; Chiadò, A.; Rivolo, P.; Geobaldo, F.; Sergo, V.; Bonifacio, A.; et al. SERS-active Ag nanoparticles on porous silicon and PDMS substrates: A comparative study of uniformity and Raman efficiency. *J. Phys. Chem. C* **2016**, *120*, 16946–16953. [CrossRef]
16. Cho, W.J.; Kim, Y.; Kim, J.K. Ultrahigh-density array of silver nanoclusters for SERS substrate with high sensitivity and excellent reproducibility. *ACS Nano* **2011**, *6*, 249–255. [CrossRef] [PubMed]
17. Liu, S.; Jiang, C.; Yang, B.; Zhang, Z.; Han, M. Controlled depositing of silver nanoparticles on flexible film and its application in ultrasensitive detection. *RSC Adv.* **2014**, *4*, 42358–42363. [CrossRef]
18. Shiohara, A.; Langer, J.; Polavarapu, L.; Liz-Marzan, L.M. Solution processed polydimethylsiloxane/gold nanostar flexible substrates for plasmonic sensing. *Nanoscale* **2014**, *6*, 9817–9823. [CrossRef] [PubMed]
19. Kvítek, L.; Prucek, R.; Panáček, A.; Novotný, R.; Hrbáč, J.; Zbořil, R. The influence of complexing agent concentration on particle size in the process of SERS active silver colloid synthesis. *J. Mater. Chem.* **2005**, *15*, 1099–1105. [CrossRef]

20. Leopold, N.; Lendl, B. A new method for fast preparation of highly surface-enhanced Raman scattering (SERS) active silver colloids at room temperature by reduction of silver nitrate with hydroxylamine hydrochloride. *J. Phys. Chem. B* **2003**, *107*, 5723–5727. [CrossRef]
21. Magdassi, S.; Grouchko, M.; Toker, D.; Kamyshny, A.; Balberg, I.; Millo, O. Ring stain effect at room temperature in silver nanoparticles yields high electrical conductivity. *Langmuir* **2005**, *21*, 10264–10267. [CrossRef] [PubMed]
22. Wang, W.; Yin, Y.; Tan, Z.; Liu, J. Coffee-ring effect-based simultaneous sers substrate fabrication and analyte enrichment for trace analysis. *Nanoscale* **2014**, *6*, 9588–9593. [CrossRef] [PubMed]
23. Xu, J.; Du, J.; Jing, C.; Zhang, Y.; Cui, J. Facile detection of polycyclic aromatic hydrocarbons by a surface-enhanced raman scattering sensor based on the au coffee ring effect. *ACS Appl. Mater. Interfaces* **2014**, *6*, 6891–6897. [CrossRef] [PubMed]
24. Tang, Y.; He, W.; Wang, S.; Tao, Z.; Cheng, L. The superiority of silver nanoellipsoids synthesized via a new approach in suppressing the coffee-ring effect during drying and film formation processes. *Nanotechnology* **2014**, *25*, 125602. [CrossRef] [PubMed]
25. Polavarapu, L.; Porta, A.L.; Novikov, S.M.; Coronado-Puchau, M.; Liz-Marzan, L.M. Pen-on-paper approach toward the design of universal surface enhanced raman scattering substrates. *Small* **2014**, *10*, 3065–3071. [CrossRef] [PubMed]
26. Cui, L.; Zhang, J.; Zhang, X.; Huang, L.; Wang, Z.; Li, Y.; Gao, H.; Zhu, S.; Wang, T.; Yang, B. Suppression of the coffee ring effect by hydrosoluble polymer additives. *ACS Appl. Mater. Interfaces* **2012**, *4*, 2775–2780. [CrossRef] [PubMed]
27. Yunker, P.J.; Still, T.; Lohr, M.A.; Yodh, A.G. Suppression of the coffee-ring effect by shape-dependent capillary interactions. *Nature* **2011**, *476*, 308–311. [CrossRef] [PubMed]
28. Mampallil, D.; Eral, H.B. A review on suppression and utilization of the coffee-ring effect. *Adv. Coll. Interface Sci.* **2018**, *252*, 38–54. [CrossRef] [PubMed]
29. Chen, R.; Zhang, L.; Li, X.; Ong, L.; Soe, Y.G.; Sinsua, N.; Gras, S.L.; Tabor, R.F.; Wang, X.; Shen, W. Trace analysis and chemical identification on cellulose nanofibers-textured SERS substrates using the "coffee ring" effect. *ACS Sens.* **2017**, *2*, 1060–1067. [CrossRef] [PubMed]
30. Zhu, J.; Chen, X.; Fan, J.; Mao, Y.; Chen, J.; Wu, W.; Jin, Y. A green chemistry synthesis of AG nanoparticles and their concentrated distribution on PDMS elastomer film for more sensitive SERS detection. In Proceedings of the 19th International Conference on Solid-State Sensors, Actuators and Microsystems (TRANSDUCERS), Kaohsiung, Taiwan, 18–22 June 2017; pp. 1395–1398.
31. Sharma, V.K.; Yngard, R.A.; Lin, Y. Silver nanoparticles: Green synthesis and their antimicrobial activities. *Adv. Coll. Interface Sci.* **2009**, *145*, 83–96. [CrossRef] [PubMed]
32. Sempels, W.; De Dier, R.; Mizuno, H.; Hofkens, J.; Vermant, J. Auto-production of biosurfactants reverses the coffee ring effect in a bacterial system. *Nat. Commun.* **2013**, *4*, 1757. [CrossRef] [PubMed]
33. Weon, B.M.; Je, J.H. Capillary force repels coffee-ring effect. *Phys. Rev. E* **2010**, *82*, 015305(R). [CrossRef] [PubMed]
34. Das, S.; Dey, A.; Reddy, G.; Sarma, D.D. Suppression of the coffee-ring effect and evaporation-driven disorder to order transition in colloidal droplets. *J. Phys. Chem. Lett.* **2017**, *8*, 4704–4709. [CrossRef] [PubMed]
35. He, A.; Yang, H.; Xue, W.; Sun, K.; Cao, Y. Tunable coffee-ring effect on a superhydrophobic surface. *Opt. Lett.* **2017**, *42*, 3936–3939. [CrossRef] [PubMed]
36. Telis, V.R.N.; Telis-Romero, J.; Mazzotti, H.B.; Gabas, A.L. Viscosity of aqueous carbohydrate solutions at different temperatures and concentrations. *Int. J. Food Prop.* **2007**, *10*, 185–195. [CrossRef]
37. Polavarapu, L.; Pérez-Juste, J.; Xu, Q.-H.; Liz-Marzán, L.M. Optical sensing of biological, chemical and ionic species through aggregation of plasmonic nanoparticles. *J. Mater. Chem. C* **2014**, *2*, 7460. [CrossRef]
38. Mao, H.; Wu, W.; She, D.; Sun, G.; Lv, P.; Xu, J. Microfluidic surface-enhanced Raman scattering sensors based on nanopillar forests realized by an oxygen-plasma-stripping-of-photoresist technique. *Small* **2014**, *10*, 127–134. [CrossRef] [PubMed]
39. Lin, D.; Wu, Z.; Li, S.; Zhao, W.; Ma, C.; Wang, J.; Jiang, Z.; Zhong, Z.; Zheng, Y.; Yang, X. Large-area au-nanoparticle-functionalized Si nanorod arrays for spatially uniform surface-enhanced Raman spectroscopy. *ACS Nano* **2017**, *11*, 1478–1487. [CrossRef] [PubMed]

40. Li, Z.; Meng, G.; Huang, Q.; Hu, X.; He, X.; Tang, H.; Wang, Z.; Li, F. Ag nanoparticle-grafted pan-nanohump array films with 3D high-density hot spots as flexible and reliable SERS substrates. *Small* **2015**, *11*, 5452–5459. [CrossRef] [PubMed]
41. Lin, J.; Shang, Y.; Li, X.; Yu, J.; Wang, X.; Guo, L. Ultrasensitive SERS detection by defect engineering on single $Cu_2O$ superstructure particle. *Adv. Mater.* **2016**, *29*. [CrossRef] [PubMed]
42. Chen, J.; Huang, Y.; Kannan, P.; Zhang, L.; Lin, Z.; Zhang, J.; Chen, T.; Guo, L. Flexible and adhesive surface enhance Raman scattering active tape for rapid detection of pesticide residues in fruits and vegetables. *Anal. Chem.* **2016**, *88*, 2149–2155. [CrossRef] [PubMed]
43. Liu, B.; Han, G.; Zhang, Z.; Liu, R.; Jiang, C.; Wang, S.; Han, M.Y. Shell thickness-dependent Raman enhancement for rapid identification and detection of pesticide residues at fruit peels. *Anal. Chem.* **2012**, *84*, 255–261. [CrossRef] [PubMed]

*nanomaterials*

MDPI

*Article*

# Steel Wire Mesh as a Thermally Resistant SERS Substrate

**Tomasz Szymborski [1,2], Evelin Witkowska [1], Krzysztof Niciński [1], Zuzanna Majka [1], Tomasz Krehlik [1], Tomiła Deskur [1], Katarzyna Winkler [1] and Agnieszka Kamińska [1,*]**

1 Institute of Physical Chemistry, Polish Academy of Sciences, Kasprzaka 44/52, Warsaw 01-224, Poland; tszymborski@gmail.com (T.S.); evelinwitkowska@wp.pl (E.W.); krzysiekn92@gmail.com (K.N.); zuza_majka@wp.pl (Z.M.); tomasz_krehlik@wp.pl (T.K.); tomila@interia.pl (T.D.); kwinkler@ichf.edu.pl (K.W.)

2 Soft Materials Laboratory, Institute of Materials, Ecole Polytechnique Fédérale de Lausanne, 1015 Lausanne, Switzerland

* Correspondence: akamin@ichf.edu.pl; Tel.: +48-22-343-3112; Fax: +48-22-343-3333

Received: 4 July 2018; Accepted: 23 August 2018; Published: 26 August 2018

**Abstract:** In this paper, we present novel type of Surface-enhanced Raman spectroscopy (SERS) platform, based on stainless steel wire mesh (SSWM) covered with thin silver layer. The stainless steel wire mesh, typically used in chemical engineering industry, is a cheap and versatile substrate for SERS platforms. SSWM consists of multiple steel wires with diameter of tens of micrometers, which gives periodical structure and high stiffness. Moreover, stainless steel provides great resistance towards organic and inorganic solvents and provides excellent heat dissipation. It is worth mentioning that continuous irradiation of the laser beam over the SERS substrate can be a source of significant increase in the local temperature of metallic nanostructures, which can lead to thermal degradation or fragmentation of the adsorbed analyte. Decomposition or fragmentation of the analysed sample usually causea a significant decrease in the intensity of recorded SERS bands, which either leads to false SERS responses or enables the analysis of spectral data. To our knowledge, we have developed for the first time the thermally resistant SERS platform. This type of SERS substrate, termed Ag/SSWM, exhibit high sensitivity (Enhancement Factor (EF) = $10^6$) and reproducibility (Relative Standard Deviation (RSD) of 6.4%) towards detection of *p*-mercaptobenzoic acid (*p*-MBA). Besides, Ag/SSWM allows the specific detection and differentiation between Gram-positive and Gram-negative bacterial species: *Escherichia coli* and *Bacillus subtilis* in label-free and reproducible manner. The unique properties of designed substrate overcome the limitations associated with photo- and thermal degradation of sensitive bacterial samples. Thus, a distinctive SERS analysis of all kinds of chemical and biological samples at high sensitivity and selectivity can be performed on the developed SERS-active substrate.

**Keywords:** Surface-enhanced Raman spectroscopy (SERS); wire mesh; steel mesh; SERS platform; *Escherichia coli; Bacillus subtilis*

## 1. Introduction

Surface-enhanced Raman spectroscopy (SERS) is nowadays one of the most powerful and reliable method in a wide range of applications including biochemistry [1], biomedical analysis [2], forensics, bio- and chemical hazards [3] and environmental monitoring [4,5].

In SERS effect, for molecules adsorbed onto metallic nanostructures, the Raman signal is amplified up to 14 orders of magnitude. Two mechanisms, an electromagnetic (EM) and a chemical (CT) are involved in SERS effect. An electromagnetic enhancement results from the amplification of incident light due to excitation of localized surface plasmon resonance (LSPR) of the metal surface. It was found that the morphology and dielectric environment of the plasmonic nanostructures plays a crucial role in the EM mechanism. The chemical enhancement process involves the charge-transfer occurring between the metal surface and adsorbed molecules, which can enhance the transition polarizability of adsorbates [6]. Typically, the electromagnetic enhancement can reach factors $10^3$–$10^{11}$, whilst the chemical enhancement contributes additional factor up to $10^3$ [7,8]. As a result of these mechanisms, SERS ensures the ultratrace detection of analytes down to single molecules [9]. Besides ultrasensitvity and fingerprint specificity, SERS technique offers nondestructive, label–free, and fast detection of analytes under a wide range of conditions (excitation wavelength and power of the laser, temperature, pressure, presence of water).

These features lead to an increase of the practical applications of SERS technique especially in biological materials studies, starting from single macromolecules and ending with whole cells (for example, microorganisms) [10–14]. The SERS spectra of biomolecules carries the information about their structure. This technique also offers the quantitative and qualitative studies with the possibility of multiplexed detection of analytes in complex fluids such as blood, cerebrospinal fluid or urine.

However, the practical implementation of SERS technique in a clinical setting is hindered by the difficulties in generation of a model SERS–active nanostructures. Such SERS substrate should reveal homogenous and high enhancement factor (EF) across the surface, as well as chemical and physical stability. Additionally, the fabrication method of SERS substrates should be as cheap as possible and provide large-scale production. The commonly used SERS substrates are based on the nanostructured noble metals e.g. Ag, Au, Cu, however the most frequently used metal is silver as it is quite cheap in comparison with gold and provides very high SERS enhancement (much higher than for copper). Interestingly, SPR (surface plasmon resonance) application of Ag is not only important from the point of view of SERS studies, but also covers other technologically important fields such as photocatalysis [15,16].

Today, a variety of techniques have been applied to fabricate SERS nanostructures, such as electrochemical methods [17], nanosphere lithography [18], electron-beam lithography [19], nanoimprinting lithography [20], vapour layer deposition [21], colloidal suspension [22], and many other strategies.

Lately, our research group has developed several types of novel nanomaterials, which can be used in a wide range of biomedical studies [23–28]. In recent research works we have investigated the membrane-based SERS-platforms [26,29], where polymer mats covered with gold nanostructures enable filtration of bacteria from fluids (for example, blood plasm), their immobilization on the filter surface, and enhancement of Raman signal. This solution overcomes the problem of transferring bacteria from filter to the SERS platform and potential contamination of the sample, and, at the same time, enables detection of very low concentration of bacteria. Nowadays, there are many different types of materials which are used to produce SERS-active substrates, e.g. ceramics [30,31], silver nanorods [32], mesoporous silica [33], ultrafiltration membrane [34], filter paper [35,36] or metal nanocrystals [37]. Particularly interesting are the polymer-based manufacturing techniques, due to the low cost and ease of production. However, polymer substrates cannot meet the demands of high stiffness of the substrate and high heat transfer (i.e. heat dissipation). Therefore, we have expanded our research by developing the thermally resistant SERS platforms. Continuous irradiation of the laser beam over the SERS substrate can be a source of significant increase in the local temperature of metallic nanostructures, what is a critical issue in SERS measurements of heat-sensitive materials [38,39]. Thermal degradation or fragmentation of the analyte is a vital problem during examination of heat-sensitive materials like DNA, proteins, polymers or lipid bilayers [40]. Moreover, molecules that are adsorbed to the metal surface (e.g. thiols at gold) can desorb at a temperature of 60–100 °C [40]. Decomposition or fragmentation of the analyte often leads to broadening and reduction of the intensity of observed spectral bands, and as a result, to misinterpretation of analytical data [41,42].

Another issue is connected with possible degradation of the polymer in contact with water (biodegradable polymers like poly(L-lactide acid) (PLLA) or polylactid acid (PLA)) or in contact with organic solvents. Biological samples involve water-based samples (blood, urine, cerebrospinal fluid), therefore used polymers should not be biodegradable in water environment, especially when the filtration process is long.

To conclude, there is a strong need for cheap, versatile, durable, stable SERS-platform which can easily dissipate heat thus can be used with high power lasers. Such platform should, preferably, be resistant to water and other organic and inorganic solvents. In this paper, we present a new type of SERS platforms based on woven wire mesh made of stainless steel (referred here as a stainless-steel wire mesh, SSWM). There are four types of woven meshes: plain weave, twilled weave, plain Dutch weave, and twilled Dutch weave, all having periodic structure which makes them excellent base for SERS platform. They are used as a filtration media in chemical industry [43,44] due to their ability to withstand large pressures, high heat resistance, small aperture and reusability. We used twilled Dutch SSWM as a base for our SERS-platform: the SSWM was cleaned and sputtered with 50 nm thick layer of silver which ensured the high amplification of the Raman signal. The dense mesh structure also provided good deposition of the analyte on the surface.

The developed SERS platforms, named Ag/SSWM, show very high surface-enhancement factor ($1 \times 10^6$), high stability (up to one month under ambient conditions), high reproducibility, and high thermal resistance to laser irradiation. These Ag/SSWM substrates demonstrated a great potential in sensitive and reproducible SERS-based detection of typical analyte like *p*-MBA, as well as whole microorganisms, that is, *Escherichia coli* and *Bacillus subtilis.*

## 2. Materials and Methods

### 2.1. Chemicals and Materials

*P*-mercaptobenzoic acid (*p*-MBA) and phosphate-buffered saline (PBS) packs (10 mM, pH = 7.2) were obtained from Sigma-Aldrich (Dorset, UK) and used without further purification. Water (resistivity over 18 MΩ), purified using a Milli-Q plus 185 system was used throughout the process. Stainless steel wire mesh was obtained with Anping County Huijin Wire Mesh Co., Ltd., Anping, China. Each type of mesh wire was purchased in quantity of 1 $m^2$ and deposited rolled in room temperature prior to use.

### 2.2. Instrumentation

#### 2.2.1. Raman and SERS Spectroscopy

Measurements were carried out with a Renishaw inVia Raman system (Wotton-under-Edge, Gloucestershire, UK) equipped with a 785 nm diode laser (Wotton-under-Edge, Gloucestershire, UK)). The light from the laser was passed through a line filter and focused on a sample mounted on an X–Y–Z translation stage with a 50× microscope objective, Numerical Aperture (NA) = 0.75. The beam diameter was approximately 5 μm. The laser power at the sample was 5 mW or less. The microscope was equipped with 1200 grooves per mm grating, cut-off optical filters, and a 1024 × 256 pixel Peltier-cooled RenCam CCD detector (Wotton-under-Edge, Gloucestershire, UK), which allowed registering the Stokes part of Raman spectra with 5–6 $cm^{-1}$ spectral resolution and 2 $cm^{-1}$ wavenumber accuracy. The experiments were performed at ambient conditions using a back-scattering geometry. The time required for completing a single SERS spectrum was 4 seconds for *p*-MBA and 40 seconds for bacteria.

The obtained spectra were processed with an OPUS software (Bruker Optic GmbH 2012 version, Ettlingen, Germany). The spectra were smoothed with Savitsky-Golay filter, the background was removed using baseline correction, and then the spectra were normalized using a so-called Min-Max normalization.

#### 2.2.2. Scanning Electron Microscopy (SEM)

Observations were performed under high vacuum using the FEI Nova NanoSEM 450 (Hillsboro, OR, USA). The accelerating voltage was in rage from 2 to 10 kV. The SSWM samples with bacteria cells were observed without any additional layer of gold.

### 2.3. *Preparation of the SERS Platform and Sample Measurement.*

General scheme of preparation of SERS-active platform for Raman measurements is shown in Figure 1. Wire mesh sample (40 × 40 mm) was cut with scissors from big sheet (100 cm × 100 cm) and placed in baker filled with acetone. In the first step the sample was sonicated for 10 min in ultrasonic bath at a temperature of 50 °C (Figure 1a). Then the acetone was exchanged and the step was repeated. After 10 min the baker was filled with isopropyl alcohol and the sample was sonicated for 10 min at a temperature of 50 °C (Figure 1b). Then the sample was sonicated for 10 min in distilled water (Millipore) at ambient temperature (Figure 1c). Cleaned wire-mesh was then dried for 30 min at 50 °C (Figure 1d) and placed in a sterile Petri dish or immediately placed in a Physical Vapour Deposition (PVD) device and sputtered with a 50 nm layer of silver (Figure 1e). The prepared SERS platform is ready for use.

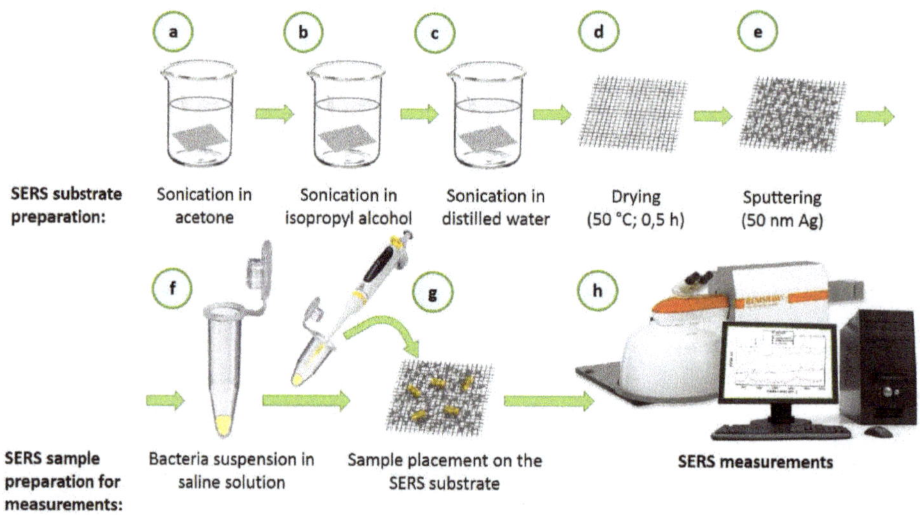


**Figure 1.** The scheme showing the preparation of SERS substrate, sample deposition and SERS measurement. The main steps involve cleaning (**a–c**), drying (**d**), sputtering of thin layer of silver (**e**), deposition of bacteria on SSWM platform (**f**,**g**), and SERS measurement (**h**).

To perform our measurements, we placed three bacterial colonies in micro-tubes filled with 500 µL of saline solution (Figure 1f), which was vortexed and then placed on platform with pipette (Figure 1g). Then the SERS platform with deposited bacteria was placed under microscope and subjected to measurements (Figure 1h).

### 2.4. *Bacteria Culture and Its Preparation for SERS Measurements*

#### 2.4.1. Bacteria Culture and SERS Sample Preparation

*E. coli* and *B. subtilis* were obtained from the Department of Applied Microbiology, Institute of Microbiology, University of Warsaw, Warsaw, Poland. In the case of each species, bacteria were streaked on the Petri dishes with LB agar (Luria-Bertani broth agar) and incubated at 37 °C for 24 h. Next, three single bacterial colonies were suspended in 500 μL of saline solution (0.9% NaCl solution) and centrifuged for 5 min at 1070× *g*. Then, the supernatant was decanted and the pellet of bacterial cells was re-suspended in 500 μL of saline solution. The obtained concentration of bacterial cells was $5 \times 10^9$ $mL^{-1}$. This process was repeated three times in order to obtain clean sample of microbial cells without any additional contamination form cell culture medium. After discarding the supernatant in the last step of centrifugation, bacteria cells were suspended in 20 μL of saline solution, transferred via pipette and placed onto SSWM substrate (Figure 1f,g). The sample was left to dry for 5 min and then the SERS measurements (Figure 1h) were conducted.

#### 2.4.2. Procedure of Silver Sputtering

To sputter a layer of silver we used the PVD equipment (EM MED020, Leica, Heerbrugg, Switzerland). The silver target was obtained from Mennica Metale Szlachetne, Warsaw, Poland. The target diameter was 54 mm, thickness was 0.5 mm, and silver purity was 4 N. The vacuum during the silver sputtering was on the level of $10^{-2}$ mbar, whereas the sputtering current was 25 mA. After the deposition process the samples were placed into a sterile Petri dish.

Six different thicknesses of silver layer on the SSWM substrates, i.e., 5, 20, 35, 50, 70 and 100 nm, were tested to find optimal conditions for SERS enhancement.

## 3. Results and Discussion

### 3.1. Characterization of Wire Mesh and Its Surface

In our experiments, we used five different wire meshes (see Table 1), made of stainless steel 316 (austenitic chromium-nickel stainless steel containing molybdenum). Addition of molybdenum increases general corrosion resistance, improves resistance to pitting from chloride ion solutions, and provides increased strength at elevated temperatures. The corrosion resistance of 316 steel is improved, particularly against acid sulphates, alkaline chlorides, and sulfuric, hydrochloric, acetic, formic and tartaric acids. What is important, the thermal conductivity of steel 316 is 16.3 $W \cdot m^{-1} \cdot K^{-1}$, whereas polymer, for example, PLA thermal conductivity is ca. 0.19 $W \cdot m^{-1} \cdot K^{-1}$ [45]. That large thermal conductivity of stainless steel is what makes it superior base material for SERS platform.

**Table 1.** Types and parameters of wire meshes used in experiments. All wire meshes are made of stainless steel 316. Mesh refers to the number of warp and weft wires per inch.

| Number of Sample | Mesh | Warp Diameter [μm] | Weft Diameter [μm] | Type of Weave |
|---|---|---|---|---|
| I | 80 × 800 | 120 | 70 | twill Dutch |
| II | 200 × 1400 | 50 | 40 | twill Dutch |
| III | 250 × 2000 | 45 | 27 | twill Dutch |
| IV | 325 × 2300 | 35 | 25 | twill Dutch |
| V | 400 × 3100 | 30 | 17 | twill Dutch |

Figure 2 presents general organization of the wires in twill Dutch weave fabric, while Figure 3 presents SEM pictures of all wires meshes (see Table 1) at three different magnifications. As can be seen, the twill Dutch type is extremely dense and tightly woven fabric. Twill refers to the structure of the fabric: over two and under two weaving wires with respect to the warp wires (warp are wires running lengthwise of the cloth, whereas weft are wires running across the cloth). The term Dutch means that the weft wires have smaller diameter than the warp wires (see Table 1 for details). The structure of twill Dutch weave is presented in Figure 3, whereas the schematic illustration of the weft and warp wires structure is presented in Figure S1.

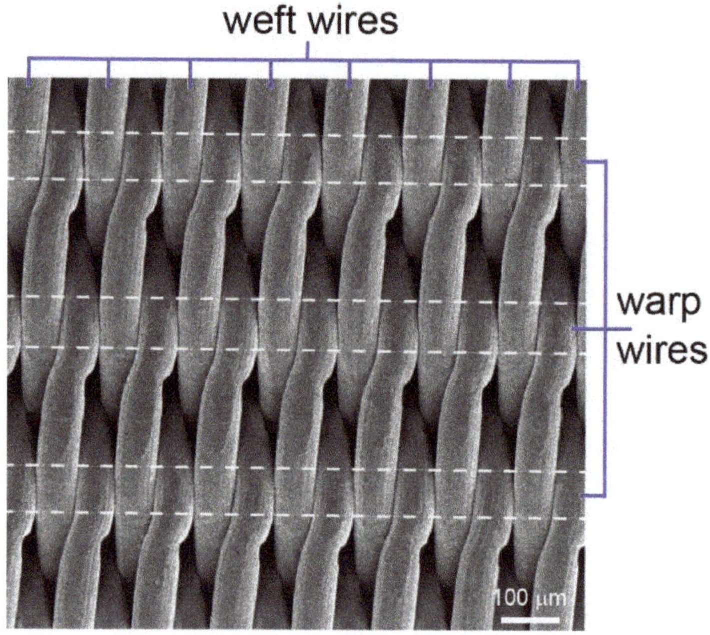


**Figure 2.** Twill Dutch weave is a densly woven fabric made of 316 stainless steel. It consists of two types of wires: warp (with a higher diameter) and weft (with a smaller diameter). The presented picture is a SEM of 80 × 800 wire mesh.

Twill Dutch weave enables the weft wires to be woven more densely. Therefore, much smaller aperture sizes can be achieved. The term mesh (e.g. 80 × 800) refers to the number of warp (80) and weft (800) wires per inch.

Figure 3 shows the representative SEM images of five wire meshes (I–V), for three different magnifications. All samples were sputtered with 50 nm of silver via PVD method as described before. This method provides a homogeneous coverage of the SSWM surface with the layer of silver.

SEM image at high magnification exhibits homogeneously placed silver nanostructures. The image analysis has been performed to quantify their size distribution. Representative image and its histogram for sample I (80 × 800) is presented in Figure 4, while the histograms for all studied samples are depicted in Figure S2. It can be noticed that for all samples the avarage size of the silver nanostructures on the surface of the wires is below 45 nm, while the median is below 40 nm. These parameters are optimal for the LSPR.

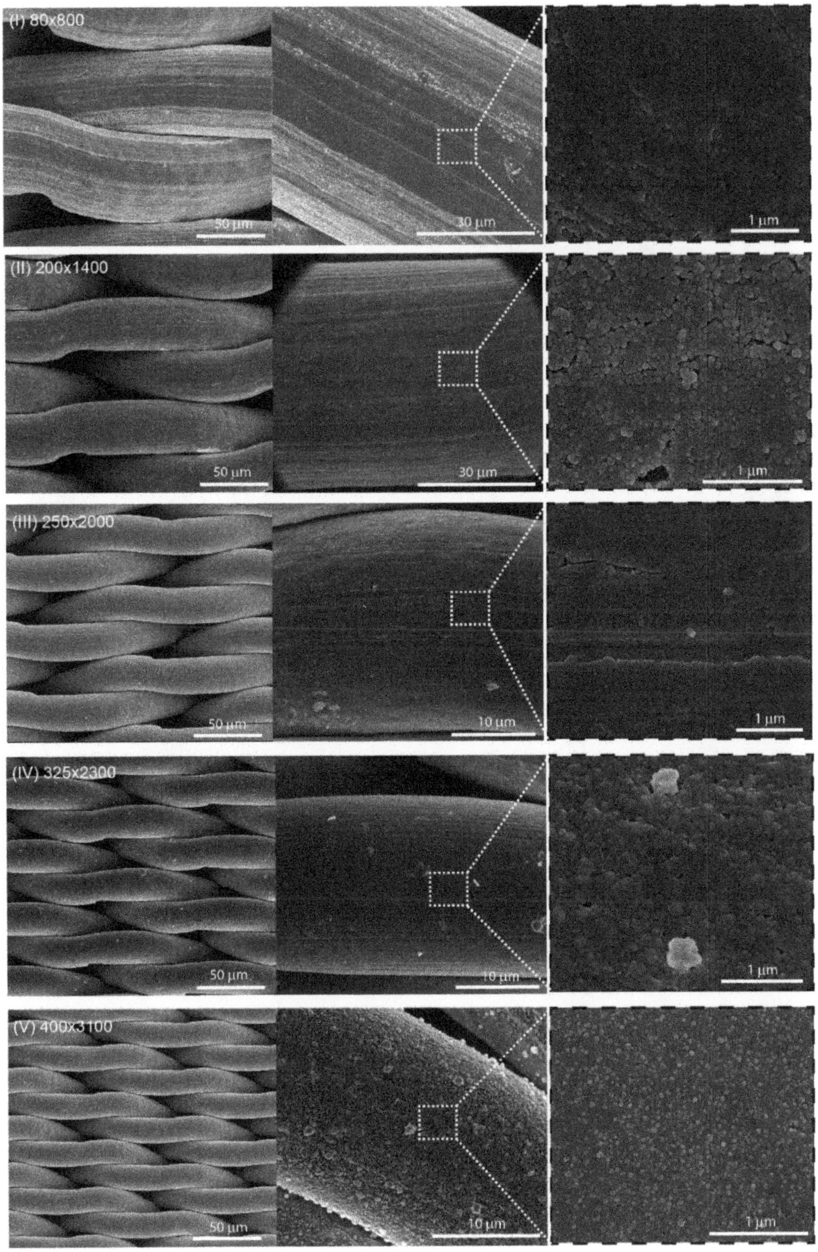


**Figure 3.** SEM images of SERS-active platforms (sputtered with 50 nm layer of silver via PVD technique) at three different magnifications. Parameters of the wire meshes are described in Table 1.

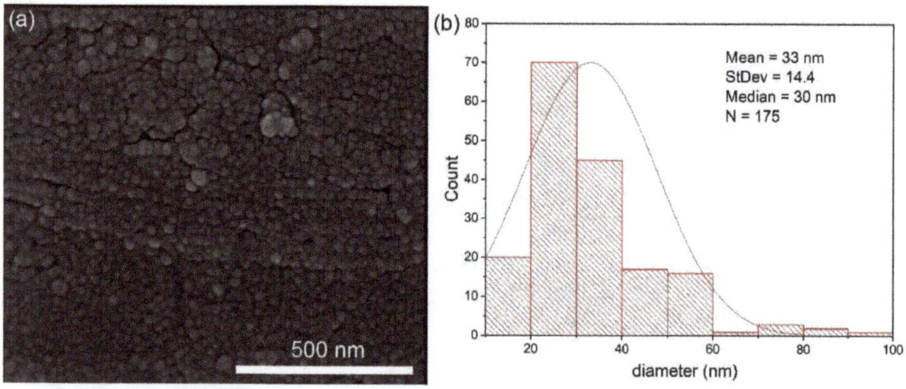


**Figure 4.** The surface of 80 × 800 (I) wire mesh covered with 50 nm layer of silver (**a**) and histogram of the size of the Ag nanostructures on the surface (**b**). The avarage size of these nanostructures is 33.0 nm ± 14.4 nm.

### 3.2. SERS Properties of Ag/SSWM Substrate

The SERS efficiency (sensitivity, selectivity, and reproducibility) is strongly correlated with the morphology of the SERS-active nanostructures. The *p*-MBA was used as a standard probe compound in order to prove the SERS properties of fabricated Ag/SSWM substrates. In this study, five different wire mesh surfaces (Figure 3) have been investigated in the terms of their spectroscopic properties.

In the first step, the SERS sensitivity of the Ag/SSWM substrates was examined and described by the enhancement factors (EF) for *p*-MBA. The *p*-MBA molecules bind through its thiol groups to the top silver layer of SERS-active surface and thus allow to record the intense SERS responses. Moreover, the intensities of *p*-MBA SERS features depend on the plasmonic properties of the SERS substrate and are not affected by the possible electronic resonance mechanism.

The Ag/SSWM substrate was kept in 1.0 mL of 1.0 × $10^{-6}$ M *p*-MBA ethanol solution for 3 h and then washed with deionized water. The Raman bands at 708, 796, 1075, 1176, 1474 and 1588 $cm^{-1}$ are characteristic for *p*-MBA [46]. The surface enhancement factors for *p*-MBA have been calculated according to the standard equation:

$$EF = \frac{I_{SERS} N_{NR}}{I_{NR} N_{SERS}} \quad (1)$$

where $N_{NR}$ stands for the number of molecules sampled by normal Raman measurements, whereas $N_{SERS}$ describes the number of molecules irradiated in SERS technique. $I_{NR}$ and $I_{SERS}$ were measured at 1075 $cm^{-1}$ and correspond to the normal Raman scattering intensity of *p*-MBA in the bulk and the SERS intensity of *p*-MBA adsorbed to metallic nanostructures.

The laser spot area and the effective illuminated volume are fundamental parameters for the estimation of EF. The effective illuminated volume has been calculated using a formula recommended by Renishaw:

$$V = 3.21 \times \lambda^3 (f/D) \quad (2)$$

where $f$ is the microscope objective focal length and D represents the effective laser beam diameter at the objective back aperture. For our setup, $V = 2012 \approx 2\times10^3$ μ$m^3$. The laser beam diameter, defined as twice the radius of a circle encompassing the area with 86% of the total power was about 2.5 μm. It should be highlighted that approximately the same values were achieved from the experimentally obtained laser spot image and from the theoretical formula ($4\lambda f/\pi D$). Assuming the volume in a shape of a cylinder with the diameter of 5 μm leads to the effective height of 100 μm. This value was confirmed by recording Raman spectra of Si while varying the distance from the focal plane.

The SERS samples were prepared by immersing the substrate in 1 mL of $1.0 \times 10^{-6}$ M solution of *p*-MBA. The number of molecules in the solution was $6.02 \times 10^{14}$ ($6.02 \times 10^{23}$ molecules/mol $\times$ 1.0 $\times 10^{-3}$ L $\times$ 1.0 $\times$ $10^{-6}$ mol/L = $6.02 \times 10^{14}$ molecules). The surface area irradiated by the laser beam (2.5 μm in diameter) was 19.6 $\mu m^2$ (3.14 $\times$ 1.25 $\mu m^2$ = 4.9 $\mu m^2$). The surface of our samples was 10 $mm^2$. Therefore, about $2.3 \times 10^8$ molecules were present in the laser beam spot. The normal Raman spectrum was observed for a cell filled with a pure *p*-MBA (154.19 $g \cdot mol^{-1}$; density of 1.06 $g \cdot cm^{-3}$). The effective illuminated volume for our setup is $2 \times 10^3$ $\mu m^3$. Under these conditions, $N_{NR} = 8.1 \times 10^{12}$ molecules were irradiated by the laser. From these data of the relative intensity and the number of molecules sampled from the regular Raman and SERS measurements, the enhancement factors for all studied SERS surfaces have been calculated.

Figure 5 depicts the SERS spectra of *p*-MBA molecules adsorbed from $10^{-6}$ M ethanol solution onto five studied surfaces (Figure 3). The enhancement factors for each surface (named from I to V) were calculated using equation (1) and are presented in Table 2. The highest enhancement factor has been found for the surface I (Table 2), which indicate that its morphology matches to the optimal size of nanostructures (20–40 nm) for the efficient LSPR.

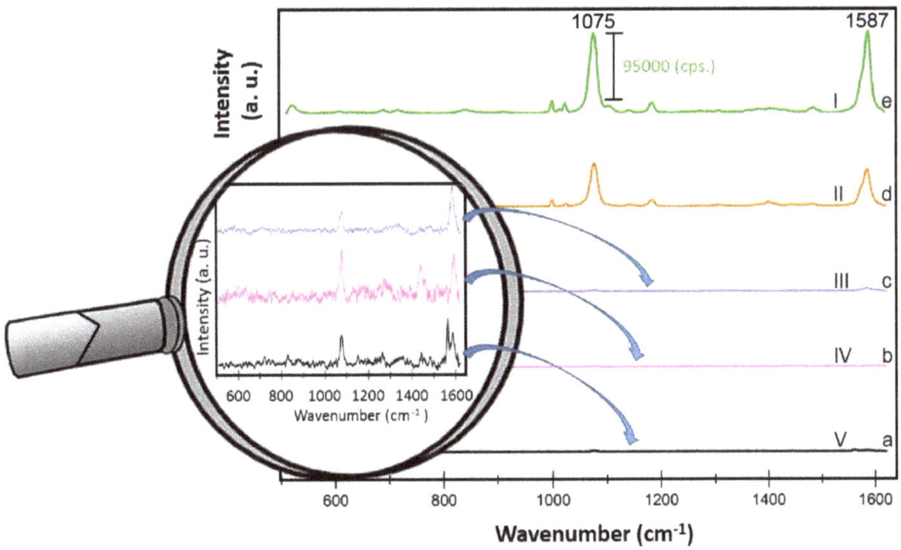


**Figure 5.** SERS spectra of *p*-MBA recorded from five different SERS substrates (a-e) with varying morphology of wire mesh (according to parameters described in Table 1). Experimental conditions: 5 mW of 785 nm excitation, 2 × 5 seconds acquisition time. The image in magnifier presents the close view of the region with marker band at 1075 $cm^{-1}$ for SSWM surfaces showing low EF. Each SERS spectrum was averaged from twenty measurements in different places of SERS surface.

**Table 2.** The EF factors for five different Ag/SSWM substrates (Table 1).

| Number | Mesh | Enhancement Factor (EF) |
|---|---|---|
| I | 80 × 800 | $4.2 \times 10^6$ |
| II | 200 × 1400 | $1.3 \times 10^5$ |
| III | 250 × 2000 | $1.0 \times 10^3$ |
| IV | 325 × 2300 | $1.2 \times 10^3$ |
| V | 400 × 3100 | $0.8 \times 10^3$ |

As was mentioned above, in order to acquire SERS activity of the SSWMs platforms, they were covered with silver layer in PVD process. The deposited silver film thickness affects the size and the density of formed silver nanostructures [47]. Figure 6 shows the dependence between the intensity of the marker bands of *p*-MBA at 1075 cm$^{-1}$ and the thickness of the deposited silver film. As can be seen the 5 nm layer of silver was insufficient to achieve the reasonable SERS responses for *p*-MBA. The intensity of *p*-MBA SERS spectrum increased significantly when the thickness of silver layer increased from 20 to 50 nm (Figure 6).

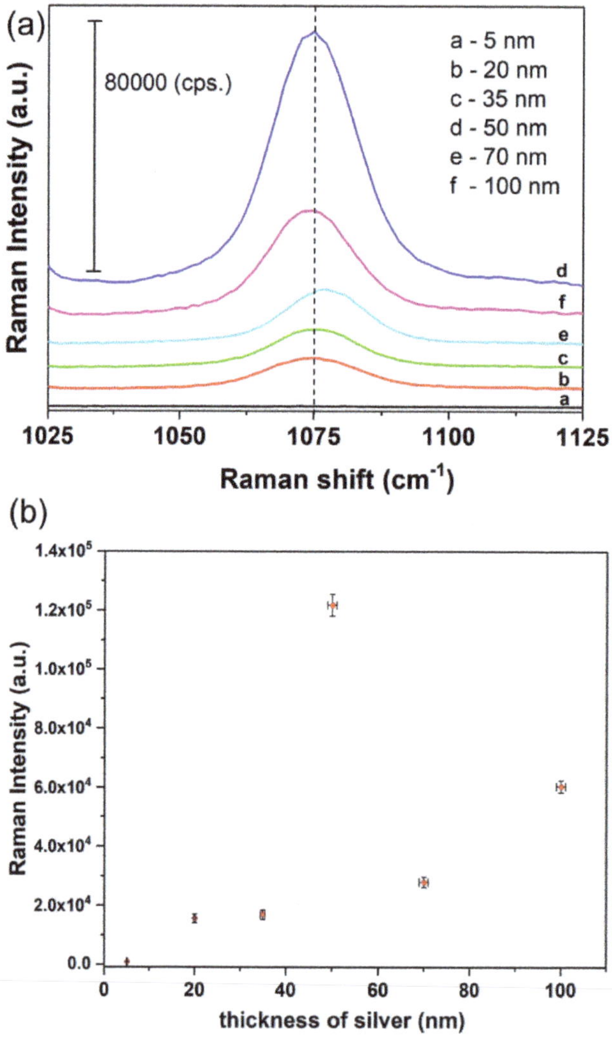


**Figure 6.** The intensity of SERS band at 1075 cm$^{-1}$ with varying Ag metal thickness (5 nm, 20 nm, 35 nm, 50 nm, 70 nm and 100 nm) for Type I surface: mesh 80 × 800: (**a**) Raman spectra, (**b**) numerical values of Raman intensity. The error bars in the Figure 6b shows standard deviation of both thickness of the silver layer and Raman signal intensity.

The maximum intensity of the marker bands of *p*-MBA has been recorded for 50 nm thickness of the deposited layers of silver. Further increase in the amount of silver leads to decrease in intensity of SERS bands.

It should be highlighted that, despite the numerous theoretical and experimental works, the nature of the Raman signal enhancement in SERS technique is still not obvious. However, as mentioned above, two enhancement mechanism (long-range electromagnetic effect (EM) and short-range charge–transfer effect) are involved in SERS phenomenon. It has been found that the dominating electromagnetic enhancement in SERS is caused by surface plasmon resonances on the substrate. This is an example of amplification of incident light intensity by excitation of surface plasmons. Briefly, the factors which determine the absorbance and the bandwidth of the plasmon resonance include the size, shape, and density of the metal nanostructures. In this case the morphology and microstructure of the Ag nanofilms onto SSWM also play a crucial role in EM effect of SERS enhancement.

These results indicate, that the 50 nm thickness of the Ag metal film results in an optimal size (average size of the objects is 33.0 nm $\pm$ 14.4 nm) and distribution of silver nanostructures for a plasmon resonance effect for 785 nm laser line. The distribution of the nanostructures determines the formation of 'hot spots' on the SERS substrate. It is possible that mentioned factors (suitable size of silver nanostructures, number and distribution of 'hot spots') make significant contribution to the observed SERS enhancement.

The thicker layers of silver (70 and 100 nm) results in decrease of SSWM roughness and thus lead to the loss of optimal condition for LSPR [48].

To summarize: (i) the stainless steel wire Type I, mesh 80 $\times$ 800 (Table 1, Figure 3) and (ii) 50 nm thickness of the Ag metal film results in obtaining the most efficient SERS-active platform. This type of substrate has been applied for further analysis of bacterial cells. In all experiments the 785 nm excitation wavelength was applied, as it is the compromise between the signal intensity and background fluorescence [49], especially during the measurements of biological samples. Additionally, we have also checked that the SERS intensity of 1075 $cm^{-1}$ marker band of *p*-MBA increases ranging from 532 to 785 nm (see Table S1 in Supplementary Materials). This behavior has been recently discussed in literature [50]. The reflectance spectra of enhancing substrates are used to the detection of far-field electromagnetic (EM) enhancement mechanisms, while the near-field modes could be responsible of the strong long-wavelength resonances in the range of 700−850 nm [50], and explain these experimental results. The dark plasmons (example of non-radiating modes) detected in nanogaps and/or aggregated nanostructures [51] are indicated as potential origins of such huge Raman signal enhancements [52].

It should be highlighted that the developed Type I SERS-active surface show excellent sensitivities for *p*-MBA, also at concentrations as low as $10^{-9}$ M. Figure S3 depicts the SERS spectra of *p*-MBA adsorbed onto Type I SERS surface at different concentrations (a) $10^{-3}$ M, (b) $10^{-6}$ M, and (c) $10^{-9}$ M in ethanol. It is clear that a steady decrease in SERS intensity of the *p*-MBA SERS band is observed with decreasing *p*-MBA concentration. However, even at $10^{-9}$ M concentration the intensity of the marker band at 1075 $cm^{-1}$ is above 1000 cps for 2 $\times$ 2 seconds of acquisition time, which still enables the SERS sensing.

### 3.3. *Applications of Ag/SSWM SERS Substrate: Differentiation Between Gram-Positive and Gram-Negative Bacteria Species*

In order to present the bioanalytical potential of developed SERS surface the SERS spectra of two different bacteria species: *E. coli* (Gram-negative) and *B. subtilis* (Gram-positive) were recorded. As can be seen in the Figure 7, the obtained SERS spectra of both bacteria species show major peaks at around 650, 725, 960, 1000, 1100, 1330, 1375, 1460 and 1590 $cm^{-1}$ and less intense bands at around 565, 780, 850, 1030, 1275 $cm^{-1}$. The peak at around 565 $cm^{-1}$ may originate from *C–O–C* ring deformation [53] or *C–C* skeletal vibration [54], while the one at 650 $cm^{-1}$—from *C–S* stretching in methionine [55] and/or *C–C* twisting mode of tyrosine [56]. The most intense band, located at 725 $cm^{-1}$, can be assigned to adenine derivatives [57]. The peak at 780 cm $^{1}$ comes probably from ring breathing modes in the DNA/RNA bases (uracil, thymine and cytosine) [58] and the one at 850 cm$^{-1}$—from asymmetric *O–P–O* stretching and/or tyrosine [59]. The bands at ~960 and 1000 cm$^{-1}$ originate probably from *C–N* stretching [60] and from phenylalanine [61], respectively. The phosphodioxy group ($PO_2^-$) from nucleic acids can be observed in the spectrum in a form of the peak at 1092 cm$^{-1}$ [62], while the amide III and $CH_2$ wagging vibrations from glycine backbone and proline sidechains can be detected as the band at 1275 cm$^{-1}$ [63]. Finally, the bands at around 1333, 1375, 1460 and 1590 cm$^{-1}$ can be assigned to adenine [64], *COO* stretching [65], $CH_2$ bending [53] and C=C olefinic stretching [60], respectively.

Although the presented in Figure 7 spectra show a lot of similar bands, there still can be noticed some peaks due to which the obtained spectra can be easily distinguished. This refers to the bands at 620, 1127, 1205, 1242 and 1402 cm$^{-1}$ which can be seen only in the case of *E. coli* spectrum, and to bands located at 1140 and 1535 cm$^{-1}$, which are detected only in *B. subtilis* spectrum. Moreover, some SERS signals, which are present in the spectra of both bacteria species, are more intense for *E. coli*, for example, 1001 and 1333 cm$^{-1}$, and some for *B. subtilis*, for example, 650 and ~1375 cm$^{-1}$. Based on these differences it can be concluded that Ag/SSWM SERS substrates can be applied for successful detection and identification of bacteria.

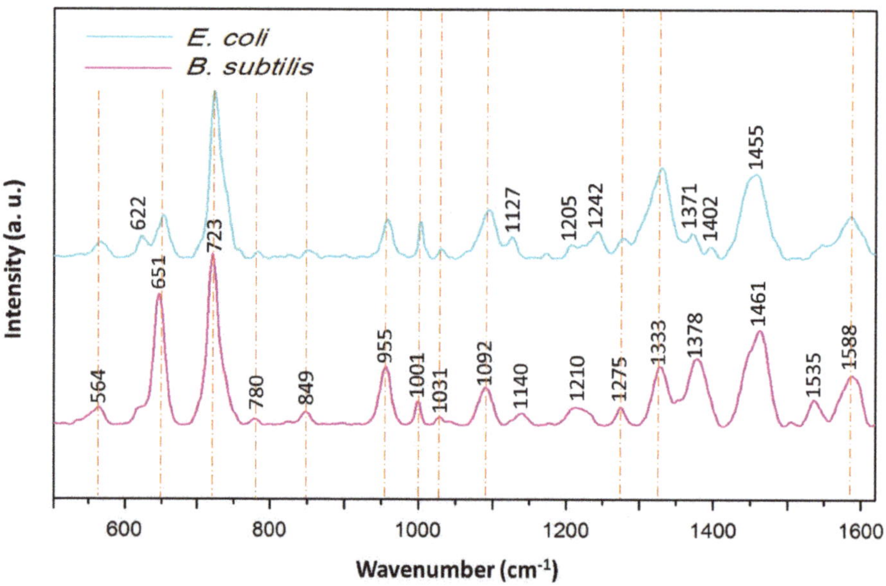


**Figure 7.** Average SERS spectra *E. coli* (turquoise) and *B. subtilis* (pink) recorded on Ag/SSWM (mesh 80 × 800 covered with 50 nm layer of Ag) SERS platforms. For all spectra, excitation wavelength was at 785 nm, laser power was 5 mW, and acquisition time was 30 seconds. Each SERS spectrum of examined bacteria was averaged from 30 measurements in different places across the SERS surface using mapping mode.

Additionally, the presence of bacteria cells onto the Ag-coated SSWM have been confirmed by SEM images (see Figure 8). The images show the *E. coli* cells, approximately 2–2.5 μm in diameter, scattered uniformly over the steel wire. Figure 8b reveals that bacteria are arranged in lines. The reason of such arrangement is connected with the structure of the stainless wire, which is produced *via* wire drawing. This process produces unevenness on the wire surface, which are parallel to the wire axis. These irregularities (scratches) are in the scale of single micrometres and perfectly match the diameter of bacteria cells.

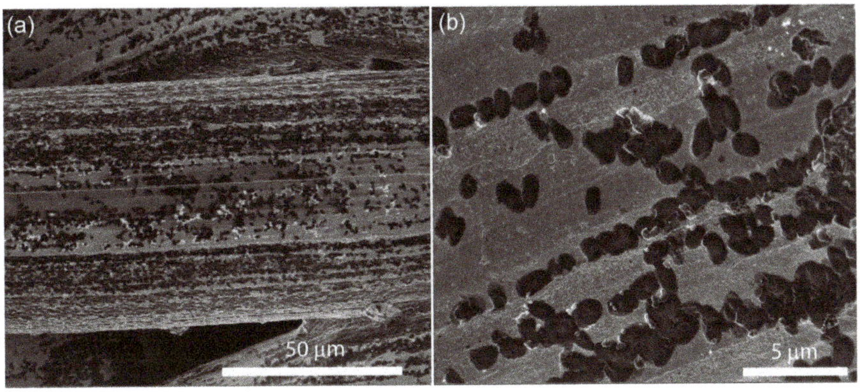


**Figure 8.** SEM images of *E. coli* placed onto Ag/SSWM surface and collected at (**a**) lower and (**b**) higher magnification. The bacteria are arranged in lines due to irregularities of wire surface.

### 3.4. RepRoducibility and Thermal Resistance of the SERS Substrate

To estimate the reproducibility of prepared Ag/SSWM substrates, the relative standard deviation (RSD) was calculated for SERS spectra of *p*-MBA, *B. subtilis*, and *E. coli*. Figures 9 and S4 presents obtained results for particular Type I substrate. The calculated RSD values equal 6.4%, 11%, and 9.5%, respectively. All these data indicate that the developed SERS substrates can be applied in biomedical and analytical studies.

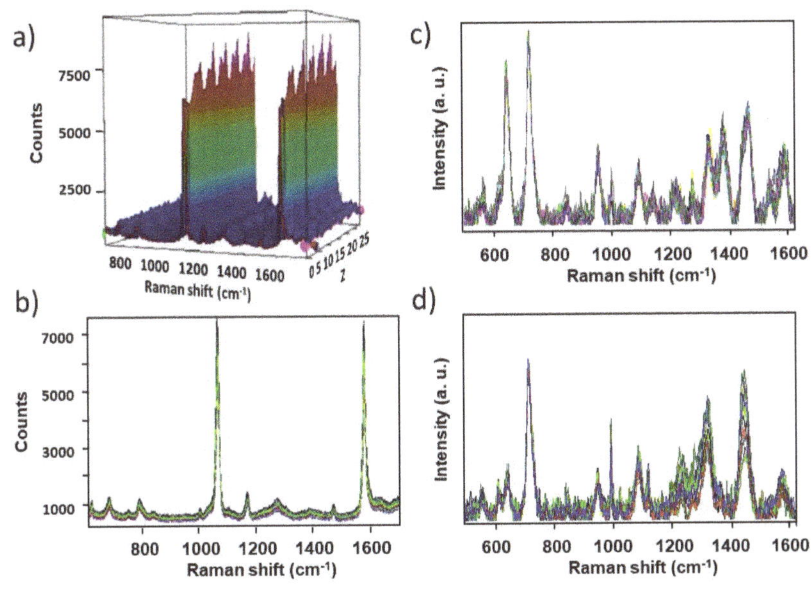


**Figure 9.** The representative SERS spectra of (**a**,**b**) *p*-MBA of concentration of $10^{-6}$ M, (**c**) *B. subtilis*, and (**d**) *E. coli* recorded from 30 different spots on the SERS surface (Type I) using mapping mode. The spectra were collected over a distance of 1 mm with 10 μm steps (30 spectra are shown). Each point in the map was recorded using 5 mW of 785 nm excitation.

The application of SERS technique within biology and medicine is a rapidly expanding field of science since SERS can provide chemical, biochemical, and structural information through the generation of fingerprint spectra or spectral imaging. This method of analysis does not require a complex sample preparation, and can also be applied without a complicated labelling strategy, and does not suffer from interference from water. However, besides the long list of advantages there are some limitations associated with SERS techniques. The wavelength of the exciting laser and its power onto the irradiated sample are the key factors that determine the spectral resolution. The spatial resolution is defined by the optic of microscope objective and the wavelength of the laser. In biological studies the NIR (near-infrared) lasers at 785 nm and 830 nm, which have relatively low photon energy and allow the reduction of fluorescence contribution in Raman spectrum, are commonly used. The power of laser illuminating the sample depends on the laser spot size and the magnification of the microscope. His in turn results in the intensity of scattered light. The biological samples are usually low-scattering materials and are very sensitive to the radiation damage or local thermal decomposition.

The burning or photo-degradation of biological samples, especially over prolonged period of excitation at too high laser power often results in appearance of the additional band in the recorded SERS spectrum at ca. 1500 cm$^{-1}$ (Figure S5). The presence of the mentioned peak is usually associated with the formation of amorphous carbon. On the other hand, the reduced laser power during measurements results in very poor spectral quality and generate an invaluable information.

To detect such week intensities of Raman scattered light and to obtain high-quality spectral features the special thermal resistant SERS-active substrates—which enable effective distribution and/or heat dissipation during intensive laser illumination—have been applied. These thermal resistant SERS-active platforms can be used not only for detection/identification of biological specimens, but also for examination of variety of thermally degradable analytes or materials.

To demonstrate the thermal capability of developed SERS-active surfaces the same volume of *E. coli* in saline solution has been deposited onto two different kinds of the SERS substrates based on: (1) poly-L-lactic acid (PLLA) and (2) SSWM surfaces. The SERS data have been recorded at various intensities of excited laser power (see Figure 10).

The results demonstrate that 14.5 mW power of laser degrades *E. coli* on PLLA polymer mat/Ag, whereas bacteria on SSWM/Ag exhibit excellent spectra. Such result can be explained by very good thermal conductance of stainless steel (nearly 100 times higher than for PLLA). Therefore, no significant increase in temperature at the measurement spot is observed in the case of SSWM. In practice, the laser power below 5 mW is used for SERS measurements of biological samples.

To summarize, the results presented in this work show an excellent sensitivity, reproducibility, and thermal stability of obtained SERS substrates (Type I), which can be successfully used during the detection of biological samples.

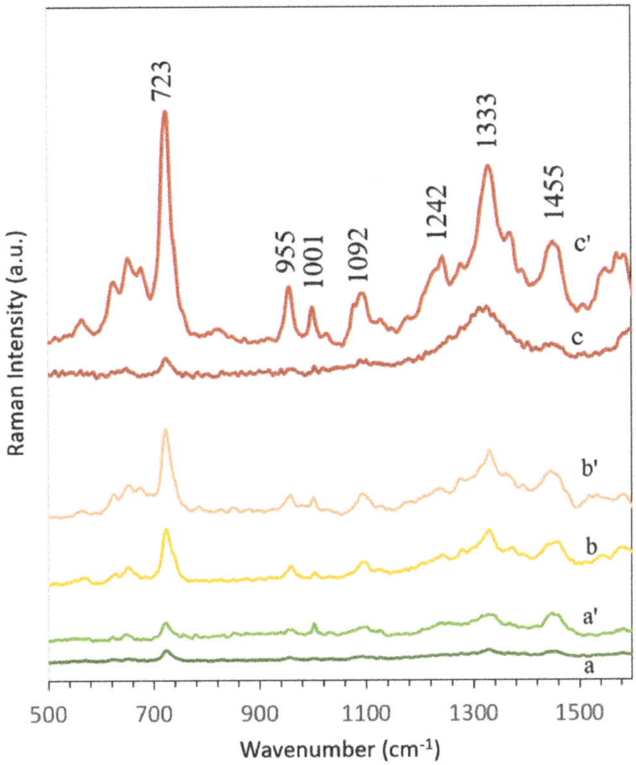


**Figure 10.** The SERS spectra of *E. coli* recorded on two different SERS substrates based on: polymer mat—PLLA (a, b, c) and the stainless steel wire mesh – SSWM (a′, b′, c′) recorded with different powers of 785 nm excitation wavelength: (a, a′) 0.6 mW, (b, b′) 1.3 mW, and (c, c′) 14.5 mW.

## 4. Conclusions

The present study demonstrates a new type of SERS platform based on woven wire mesh made of stainless steel (SSWM). The influence of the diameter of wires, their morphology and metal thickness on SERS efficiency was investigated and the optimal fabrication process parameters were established. In this paper, for the first time, we present the SERS substrate that exhibits excellent heat dissipation, which is a critical factor in SERS measurements of thermally sensitive samples. Our substrate enables generation of the high spectral resolution data even for very poorly scattering biological materials *via* safe adjustment of laser intensity to higher values. This approach overcomes the limitations associated with photo-and thermal degradation of sensitive materials, improves the spectral intensities of week Raman scatterers, and thus extends the rage of valuable SERS applications. In addition, the developed SERS substrate enables the detection and differentiation of Gram-positive and Gram-negative bacterial species in label-free manner, based on their high-quality spectral features.

**Supplementary Materials:** The following are available online at http://www.mdpi.com/2079-4991/8/9/663/s1.

**Author Contributions:** Conceptualization, T.S. and A.K., Data curation, E.W., Investigation, Z.M. and T.K and K.W., Methodology, K.N Validation, T.D., Writing original draft, A.K.

**Funding:** Foundation for Polish Science under grant Team-Tech/2017-4/23.

**Conflicts of Interest:** The authors declare no conflict of interest.

## References


1. Wilson, C.G.; Graham, D.; McBride, E.; Girkin, J.M.; Stokes, R.J.; Smith, W.E. Surface-Enhanced Raman Scattering Spectroscopy as a Sensitive and Selective Technique for the Detection of Folic Acid in Water and Human Serum. *Appl. Spectrosc.* **2008**, *62*, 371–376.
2. Kneipp, K.; Kneipp, H.; Kartha, V.B.; Manoharan, R.; Deinum, G.; Itzkan, I.; Dasari, R.R.; Feld, M.S. Detection and identification of a single DNA base molecule using surface-enhanced Raman scattering (SERS). *Phys. Rev. E* **1998**, *57*, R6281–R6284. [CrossRef]
3. Rodger, C.; Dent, G.; Watkinson, J.; Smith, W.E. Surface-Enhanced Resonance Raman Scattering and Near-Infrared Fourier Transform Raman Scattering as *in Situ* Probes of Ink Jet Dyes Printed on Paper. *Appl. Spectrosc.* **2000**, *54*, 1567–1576. [CrossRef]
4. Faulds, K.; Smith, W.E.; Graham, D.; Lacey, R.J. Assessment of silver and gold substrates for the detection of amphetamine sulfate by surface enhanced Raman scattering (SERS). *Analyst* **2002**, *127*, 282–286. [CrossRef] [PubMed]
5. Koglin, E.; Séquaris, J.-M. *Surface Enhanced Raman Scattering of Biomolecules*; Springer: Berlin/Heidelberg, Germany, 1986; pp. 1–57.
6. Campion, A.; Kambhampati, P. Surface-enhanced Raman scattering. *Chem. Soc. Rev.* **1998**, *27*, 241. [CrossRef]
7. Stiles, P.L.; Dieringer, J.A.; Shah, N.C.; Van Duyne, R.P. Surface-Enhanced Raman Spectroscopy. *Annu. Rev. Anal. Chem.* **2008**, *1*, 601–626. [CrossRef] [PubMed]
8. Camden, J.P.; Dieringer, J.A.; Wang, Y.; Masiello, D.J.; Marks, L.D.; Schatz, G.C.; Van Duyne, R.P. Probing the Structure of Single-Molecule Surface-Enhanced Raman Scattering Hot Spots. *J. Am. Chem. Soc.* **2008**, *130*, 12616–12617. [CrossRef] [PubMed]
9. Le Ru, E.C.; Etchegoin, P.G. Single-Molecule Surface-Enhanced Raman Spectroscopy. *Annu. Rev. Phys. Chem.* **2012**, *63*, 65–87. [CrossRef] [PubMed]
10. Sivanesan, A.; Witkowska, E.; Adamkiewicz, W.; Dziewit, Ł.; Kamińska, A.; Waluk, J. Nanostructured silver–gold bimetallic SERS substrates for selective identification of bacteria in human blood. *Analyst* **2014**, *139*, 1037. [CrossRef] [PubMed]
11. Alexander, T.A.; Le, D.M. Characterization of a commercialized SERS-active substrate and its application to the identification of intact Bacillus endospores. *Appl. Opt.* **2007**, *46*, 3878–3890. [CrossRef] [PubMed]
12. Bonifacio, A.; Millo, D.; Gooijer, C.; Boegschoten, R.; van der Zwan, G. Linearly moving low-volume spectroelectrochemical cell for microliter-scale surface-enhanced resonance Raman spectroscopy of heme proteins. *Anal. Chem.* **2004**, *76*, 1529–1531. [CrossRef] [PubMed]
13. Nabiev, I.R.; Morjani, H.; Manfait, M. Selective analysis of antitumor drug interaction with living cancer cells as probed by surface-enhanced Raman spectroscopy. *Eur. Biophys. J.* **1991**, *19*, 311–316. [CrossRef] [PubMed]
14. Morjani, H.; Riou, J.F.; Nabiev, I.; Lavelle, F.; Manfait, M. Molecular and cellular interactions between intoplicine, DNA, and topoisomerase II studied by surface-enhanced Raman scattering spectroscopy. *Cancer Res.* **1993**, *53*, 4784–4790. [PubMed]
15. Chen, K.-H.; Pu, Y.-C.; Chang, K.-D.; Liang, Y.-F.; Liu, C.-M.; Yeh, J.-W.; Shih, H.-C.; Hsu, Y.-J. Ag-Nanoparticle-Decorated $SiO_2$ Nanospheres Exhibiting Remarkable Plasmon-Mediated Photocatalytic Properties. *J. Phys. Chem. C* **2012**, *116*, 19039–19045. [CrossRef]
16. Chen, W.-T.; Hsu, Y.-J.; Kamat, P.V. Realizing Visible Photoactivity of Metal Nanoparticles: Excited-State Behavior and Electron-Transfer Properties of Silver ($Ag_8$) Clusters. *J. Phys. Chem. Lett.* **2012**, *3*, 2493–2499. [CrossRef] [PubMed]
17. Wang, H.-H.; Liu, C.-Y.; Wu, S.-B.; Liu, N.-W.; Peng, C.-Y.; Chan, T.-H.; Hsu, C.-F.; Wang, J.-K.; Wang, Y.-L. Highly Raman-Enhancing Substrates Based on Silver Nanoparticle Arrays with Tunable Sub-10nm Gaps. *Adv. Mater.* **2006**, *18*, 491–495. [CrossRef]
18. Dick, L.A.; McFarland, A.D.; Haynes, C.L.; Van Duyne, R.P. Metal film over nanosphere (MFON) electrodes for surface-enhanced Raman spectroscopy (SERS): Improvements in surface nanostructure stability and suppression of irreversible loss. *J. Phys. Chem. B* **2001**, *106*, 853–860. [CrossRef]
19. Fromm, D.P.; Sundaramurthy, A.; Kinkhabwala, A.; Schuck, P.J.; Kino, G.S.; Moerner, W.E. Exploring the chemical enhancement for surface-enhanced Raman scattering with Au bowtie nanoantennas. *J. Chem. Phys.* **2006**, *124*, 61101. [CrossRef] [PubMed]

20. Alvarez-Puebla, R.; Cui, B.; Bravo-Vasquez, J.-P.; Veres, T. Nanoimprinted SERS-Active Substrates with Tunable Surface Plasmon Resonances. *J. Phys. Chem. C* **2007**, *111*, 6720–6723. [CrossRef]
21. Semin, D.J.; Rowlen, K.L. Influence of Vapor Deposition Parameters on SERS Active Ag Film Morphology and Optical Properties. *Anal. Chem* **1994**, *66*, 4324–4331. [CrossRef]
22. Faulds, K.; Smith, W.E.; Graham, D. Evaluation of Surface-Enhanced Resonance Raman Scattering for Quantitative DNA Analysis. *Anal. Chem.* **2004**, *76*, 412–417. [CrossRef] [PubMed]
23. Kamińska, A.; Witkowska, E.; Kowalska, A.; Skoczyńska, A.; Ronkiewicz, P.; Szymborski, T.; Waluk, J. Rapid detection and identification of bacterial meningitis pathogens in: Ex vivo clinical samples by SERS method and principal component analysis. *Anal. Methods* **2016**, *8*. [CrossRef]
24. Kamińska, A.; Sprynskyy, M.; Winkler, K.; Szymborski, T. Ultrasensitive SERS immunoassay based on diatom biosilica for detection of interleukins in blood plasma. *Anal. Bioanal. Chem.* **2017**, *409*, 6337–6347. [CrossRef] [PubMed]
25. Kamińska, A.; Winkler, K.; Kowalska, A.; Witkowska, E.; Szymborski, T.; Janeczek, A.; Waluk, J. SERS-based Immunoassay in a Microfluidic System for the Multiplexed Recognition of Interleukins from Blood Plasma: Towards Picogram Detection. *Sci. Rep.* **2017**, *7*. [CrossRef] [PubMed]
26. Witkowska, E.; Szymborski, T.; Kamińska, A.; Waluk, J. Polymer mat prepared via Forcespinning$^{TM}$ as a SERS platform for immobilization and detection of bacteria from blood plasma. *Mater. Sci. Eng. C* **2017**, *71*. [CrossRef] [PubMed]
27. Kamińska, A.; Szymborski, T.; Jaroch, T.; Zmysłowski, A.; Szterk, A. Gold-capped silicon for ultrasensitive SERS-biosensing: Towards human biofluids analysis. *Mater. Sci. Eng. C* **2018**, *84*, 208–217. [CrossRef] [PubMed]
28. Kamińska, A.; Witkowska, E.; Winkler, K.; Dzięcielewski, I.; Weyher, J.L.; Waluk, J. Detection of Hepatitis B virus antigen from human blood: SERS immunoassay in a microfluidic system. *Biosens. Bioelectron.* **2015**, *66*, 461–467. [CrossRef] [PubMed]
29. Szymborski, T.; Witkowska, E.; Adamkiewicz, W.; Waluk, J.; Kamińska, A. Electrospun polymer mat as a SERS platform for the immobilization and detection of bacteria from fluids. *Analyst* **2014**, *139*. [CrossRef] [PubMed]
30. Mosier-Boss, P.A.; Sorensen, K.C.; George, R.D.; Sims, P.C.; O'braztsova, A. SERS substrates fabricated using ceramic filters for the detection of bacteria: Eliminating the citrate interference. *Spectrochim. Acta Part A Mol. Biomol. Spectrosc.* **2017**, *180*, 161–167. [CrossRef] [PubMed]
31. Mosier-Boss, P.A.; Sorensen, K.C.; George, R.D.; Obraztsova, A. SERS substrates fabricated using ceramic filters for the detection of bacteria. *Spectrochim Acta A Mol Biomol Spectrosc.* **2016**, *15*, 591–598. [CrossRef] [PubMed]
32. Chen, J.; Wu, X.; Huang, Y.-W.; Zhao, Y. Detection of E. coli using SERS active filters with silver nanorod array. *Sensors Actuators B. Chem.* **2014**, *191*, 485–490. [CrossRef]
33. Lin, C.-C.; Yang, Y.-M.; Liao, P.-H.; Chen, D.-W.; Lin, H.-P.; Chang, H.-C. A filter-like AuNPs@MS SERS substrate for Staphylococcus aureus detection. *Biosens. Bioelectron.* **2013**, *53*, 519–527. [CrossRef] [PubMed]
34. Hong, J.; Kawashima, A.; Hamada, N. A simple fabrication of plasmonic surface-enhanced Raman scattering (SERS) substrate for pesticide analysis via the immobilization of gold nanoparticles on UF membrane. *Appl. Surf. Sci.* **2017**, *407*, 440–446. [CrossRef]
35. Rapid fabrication of silver nanoparticle-coated filter paper as SERS substrate for low-abundance molecules detection. *Spectrochim. Acta Part A Mol. Biomol. Spectrosc.* **2017**, *179*, 211–215. [CrossRef] [PubMed]
36. In situ silver nanoparticles synthesis in agarose film supported on filter paper and its application as highly efficient SERS test stripes. *Forensic Sci. Int.* **2014**, *237*, e42–e46. [CrossRef] [PubMed]
37. Lin, W.-H.; Lu, Y.-H.; Hsu, Y.-J. Au nanoplates as robust, recyclable SERS substrates for ultrasensitive chemical sensing. *J. Colloid Interface Sci.* **2014**, *418*, 87–94. [CrossRef] [PubMed]
38. Honda, M.; Saito, Y.; Smith, N.I.; Fujita, K.; Kawata, S. Nanoscale heating of laser irradiated single gold nanoparticles in liquid. *Opt. Express* **2011**, *19*, 12375. [CrossRef] [PubMed]
39. Qin, Z.; Bischof, J.C. Thermophysical and biological responses of gold nanoparticle laser heating. *Chem. Soc. Rev.* **2012**, *41*, 1191–1217. [CrossRef] [PubMed]
40. Mochizuki, M.; Asatyas, S.; Suthiwanich, K.; Hayashi, T. Thiol Molecules as Temperature Sensors for Surface-enhanced Raman Scattering Measurements of Heat-sensitive Materials. *Chem. Lett.* **2016**, *45*, 1207–1209. [CrossRef]

41. De Jesús, M.A.; Giesfeldt, K.S.; Sepaniak, M.J. Use of a Sample Translation Technique to Minimize Adverse Effects of Laser Irradiation in Surface-Enhanced Raman Spectrometry. *Appl. Spectrosc.* **2003**, *57*, 428–438. [CrossRef] [PubMed]
42. Suh, J.S.; Jeong, D.H.; Lee, M.S. Effect of inhomogeneous broadening on the surface photochemistry of phthalazine. *J. Raman Spectrosc.* **1999**, *30*, 595–598. [CrossRef]
43. Song, B. Simple and fast fabrication of superhydrophobic metal wire mesh for efficiently gravity-driven oil/water separation. *Mar. Pollut. Bull.* **2016**. [CrossRef] [PubMed]
44. Yoshida, Y.; Inoue, Y.; Shimosaka, A.; Shirakawa, Y.; Hidaka, J. Effect of aperture structure of Dutch weave mesh on flow resistivity. *J. Chem. Eng. Japan* **2015**. [CrossRef]
45. Lebedev, S.M.; Gefle, O.S.; Amitov, E.T.; Berchuk, D.Y.; Zhuravlev, D.V. Poly(lactic acid)-based polymer composites with high electric and thermal conductivity and their characterization. *Polym. Test.* **2017**, *58*, 241–248. [CrossRef]
46. Michota, A.; Bukowska, J. Surface-enhanced Raman scattering (SERS) of 4-mercaptobenzoic acid on silver and gold substrates. *J. Raman Spectrosc.* **2003**, *34*, 21–25. [CrossRef]
47. Fang, Y.; Seong, N.-H.; Dlott, D.D. Measurement of the distribution of site enhancements in surface-enhanced Raman scattering. *Science* **2008**, *321*, 388–392. [CrossRef] [PubMed]
48. Terekhov, S.N.; Kachan, S.M.; Panarin, A.Y.; Mojzes, P. Surface-enhanced Raman scattering on silvered porous alumina templates: role of multipolar surface plasmon resonant modes. *Phys. Chem. Chem. Phys.* **2015**, *17*, 31780–31789. [CrossRef] [PubMed]
49. Petrou, V.; Hill, W. Caries Detection by Diode Laser Raman Spectroscopy. *Appl. Spectrosc.* **2000**, *54*, 795–799.
50. Novara, C.; Dalla Marta, S.; Virga, A.; Lamberti, A.; Angelini, A.; Chiadò, A.; Rivolo, P.; Geobaldo, F.; Sergo, V.; Bonifacio, A.; et al. SERS-Active Ag Nanoparticles on Porous Silicon and PDMS Substrates: A Comparative Study of Uniformity and Raman Efficiency. *J. Phys. Chem. C* **2016**, *120*, 16946–16953. [CrossRef]
51. Herzog, J.B.; Knight, M.W.; Li, Y.; Evans, K.M.; Halas, N.J.; Natelson, D. Dark Plasmons in Hot Spot Generation and Polarization in Interelectrode Nanoscale Junctions. *Nano Lett.* **2013**, *13*, 1359–1364. [CrossRef] [PubMed]
52. Hastings, S.P.; Swanglap, P.; Qian, Z.; Fang, Y.; Park, S.-J.; Link, S.; Engheta, N.; Fakhraai, Z. Quadrupole-Enhanced Raman Scattering. *ACS Nano* **2014**, *8*, 9025–9034. [CrossRef] [PubMed]
53. Luna-Pineda, T.; Soto-Feliciano, K.; De La Cruz-Montoya, E.; Pacheco Londoño, L.C.; Ríos-Velázquez, C.; Hernández-Rivera, S.P. Spectroscopic characterization of biological agents using FTIR, normal Raman and surface-enhanced Raman spectroscopies. In Proceedings of the SPIE 6554, Chemical and Biological Sensing VIII, Orlando, FL, USA, 27 April 2007; Fountain III, A.W., Ed.; Volume 6554, p. 65540K.
54. Pearman, W.F.; Lawrence-Snyder, M.; Angel, S.M.; Decho, A.W. Surface-Enhanced Raman Spectroscopy for *in Situ* Measurements of Signaling Molecules (Autoinducers) Relevant to Bacteria Quorum Sensing. *Appl. Spectrosc.* **2007**, *61*, 1295–1300. [CrossRef] [PubMed]
55. Guicheteau, J.; Argue, L.; Emge, D.; Hyre, A.; Jacobson, M.; Christesen, S. *Bacillus* Spore Classification via Surface-Enhanced Raman Spectroscopy and Principal Component Analysis. *Appl. Spectrosc.* **2008**, *62*, 267–272. [CrossRef] [PubMed]
56. Chao, Y.; Zhang, T. Surface-enhanced Raman scattering (SERS) revealing chemical variation during biofilm formation: from initial attachment to mature biofilm. *Anal. Bioanal. Chem.* **2012**, *404*, 1465–1475. [CrossRef] [PubMed]
57. Kubryk, P.; Niessner, R.; Ivleva, N.P. The origin of the band at around 730 $cm^{-1}$ in the SERS spectra of bacteria: a stable isotope approach. *Analyst* **2016**, *141*, 2874–2878. [CrossRef] [PubMed]
58. Madzharova, F.; Heiner, Z.; Gühlke, M.; Kneipp, J. Surface-Enhanced Hyper-Raman Spectra of Adenine, Guanine, Cytosine, Thymine, and Uracil. *J. Phys. Chem. C* **2016**, *120*, 15415–15423. [CrossRef] [PubMed]
59. Movasaghi, Z.; Rehman, S.; Rehman, I.U. Raman Spectroscopy of Biological Tissues. *Appl. Spectrosc. Rev.* **2007**, *42*, 493–541. [CrossRef]
60. Laucks, M.L.; Sengupta, A.; Junge, K.; Davis, E.J.; Swanson, B.D. Comparison of Psychro-Active Arctic Marine Bacteria and Common Mesophillic Bacteria Using Surface-Enhanced Raman Spectroscopy. *Appl. Spectrosc.* **2005**, *59*, 1222–1228. [CrossRef] [PubMed]
61. Zhang, J.; Huang, Q.; Yao, G.; Ke, Z.; Zhang, H.; Lu, Y. SERS study of transformation of phenylalanine to tyrosine under particle irradiation. *J. Mol. Struct.* **2014**, *1072*, 195–202. [CrossRef]

62. Krafft, C.; Neudert, L.; Simat, T.; Salzer, R. Near infrared Raman spectra of human brain lipids. *Spectrochim. Acta Part A Mol. Biomol. Spectrosc.* **2005**, *61*, 1529–1535. [CrossRef] [PubMed]
63. Dukor, R.K. Vibrational Spectroscopy in the Detection of Cancer. In *Handbook of Vibrational Spectroscopy*; Griffiths, P.R., Ed.; John Wiley & Sons, Ltd.: Chichester, UK, 2006; ISBN 9780470027325.
64. Walter, A.; März, A.; Schumacher, W.; Rösch, P.; Popp, J. Towards a fast, high specific and reliable discrimination of bacteria on strain level by means of SERS in a microfluidic device. *Lab Chip* **2011**, *11*, 1013. [CrossRef] [PubMed]
65. Demirel, M.C.; Kao, P.; Malvadkar, N.; Wang, H.; Gong, X.; Poss, M.; Allara, D.L. Bio-organism sensing via surface enhanced Raman spectroscopy on controlled metal/polymer nanostructured substrates. *Biointerphases* **2009**, *4*, 35–41. [CrossRef] [PubMed]

nanomaterials

MDPI

Article

# Rapid Tartrazine Determination in Large Yellow Croaker with Ag Nanowires Using Surface-Enhanced Raman Spectroscopy

**Jia Song [1], Yuanyi Zhang [1], Yiqun Huang [2], Yuxia Fan [1,3] and Keqiang Lai [1,3,*]**

1 College of Food Science and Technology, Shanghai Ocean University, Shanghai 201306, China; susanjia123@163.com (J.S.); jennifer.yuanyizhang@foxmail.com (Y.Z.); yxfan@shou.edu.cn (Y.F.)
2 School of Chemistry & Biological Engineering, Changsha University of Science & Technology, Changsha 410076, Hunan, China; yiqunh@csust.edu.cn
3 Engineering Research Center of Food Thermal Processing Technology, Shanghai Ocean University, Shanghai 201306, China
* Correspondence: kqlai@shou.edu.cn; Tel.: +86-21-6190-0754

Received: 1 November 2018; Accepted: 17 November 2018; Published: 23 November 2018

**Abstract:** In this work, surface-enhanced Raman spectroscopy (SERS) technology coupled with Ag nanowires was shown to be a promising tool in the detection of tartrazine in large yellow croaker for the first time. Ag nanowires with a uniform diameter were fabricated by an efficient and manageable polyol method. The partial least square model was established for the quantitative analysis of tartrazine, which showed a relatively high linear correlation between actual and predicted concentrations of standard tartrazine solutions. An optimal sample preparation method was also selected and used to extract tartrazine from large yellow croaker within 20 min. The lowest concentration detected was 20.38 ng/cm$^2$, which fully meets the requirements of tartrazine testing in aquatic products. This study indicated that SERS technology combined with the as-prepared Ag nanowires could detect tartrazine sensitively and provide an easily operable and time-saving way to monitor tartrazine in large yellow croaker.

**Keywords:** SERS; Ag nanowires; tartrazine; large yellow croaker

## 1. Introduction

Tartrazine is a well-known synthetic colorant that is frequently and widely used in foods (such as soft drinks and candies), pharmaceuticals and cosmetics to make them appealing to customers [1]. The acceptable daily intake of tartrazine in food, proposed by the FAO and WHO, is 7.5 mg/kg. Tartrazine has low toxicity, but its excessive ingestion can result in allergy, asthma, migraine and hyperactivity [2]. It may also trigger sleep disorders in children [3]. Given the potential risks to humans, tartrazine has been banned in countries including Norway and Austria [4]. Increasing concern has been paid to health issues provoked by food additives used in China; the types of foodstuff and the maximum amount of additives allowed have been strictly set by Chinese regulation (GB2760-2014) [5]. Tartrazine, a coloring agent used on yellow croaker (*Pseudosciaena crocea*), in tsukemono and in wine, has been banned by the Ministry of Health of the Republic of China [6]. However, it is still used illegally and in excess in various food products due to its high economic benefits. Hence, developing a time-efficient, cost-effective and sensitive method to detect tartrazine in foodstuff is necessary.

Currently, numerous methods have been developed to analyze tartrazine in different foods. These methods include high-performance liquid chromatography (HPLC) [7], liquid chromatography-tandem quadrupole mass spectrometry [8], absorption spectrometry [9], voltammetry [10] and thin-layer chromatography [11]. Although some of these analytical methods have high accuracy, good reproducibility and reach the necessary level of quantification according to regulations, they also have disadvantages, such as high cost, inconvenient operation, and complicated sample preparation. Surface-enhanced Raman scattering or surface-enhanced Raman spectroscopy (SERS) is one of the most effective analytical tools for detecting the trace components of food owing to its high sensitivity, quick determination and convenient operation. A growing number of research publications on the detection of illegal additives [12], pesticide residues [13–16] and veterinary drugs in foodstuff are available [17,18]. There are some reports on the use of SERS for the detection of tartrazine [19–21] but the number of publications on the determination of tartrazine in foodstuff on the basis of the SERS method is relatively low. In 2016, SERS coupled with Au nanodumbbells was used to analyze tartrazine in drinks [22].

To the best of our knowledge, no report using SERS to detect tartrazine in large yellow croaker has been published. The main objective of this study is to investigate and develop a time-efficient and cost-effective method of detecting trace amounts of tartrazine in large yellow croaker using Ag nanowires.

## 2. Materials and Methods

### 2.1. Materials and Reagents

$AgNO_3$ (>99%), polyvinyl pyrrolidone (PVP) (Mw = 55,000 g/mol) and NaCl (≥99.999%) were bought from Sigma-Aldrich (St. Louis, MO, USA). Glycerol (American Chemical Society grade) was obtained from Aladdin Industrial Corporation (Shanghai, China). Ethanol (analytical grade) was received from Sinopharm Chemical Reagent Co., Ltd. (Shanghai, China). Tartrazine (≥99.0%) and methanol (HPLC grade) were purchased from Sigma-Aldrich (St. Louis, MO, USA). $MgSO_4$ was purchased from Aladdin Industrial Corporation (Shanghai, China). Large yellow croaker were bought from a supermarket in Shanghai and we confirmed an absence of tartrazine by the HPLC method, which was performed in the China General Chamber of Commerce Supervision & Testing Center for Food (Shanghai, China). The 0.45 μm polytetrafluoroethene (PTFE) microporous films were obtained from Anpel Corporation (Shanghai, China).

### 2.2. Preparation and Characterization of Ag Nanowires

Ag nanowires were synthesized using the polyol method referred to in our previously published article [23]. Ag nanowires were fabricated by reducing $AgNO_3$ in glycerol at a rising temperature. PVP, along with the mixture of NaCl, ultrapure water and glycerol, was also added to the reaction liquid; PVP was the stabilizing agent or surfactant. Then, the cooled products were washed by ultrapure water and ethanol aqueous solution. The UV-Vis absorbance spectrum of the as-fabricated Ag nanowires was recorded by a UV-Vis spectrometer (UV3000PC, MAPADA Instruments Ltd., Shanghai, China). Transmission electron microscopy (TEM; JEM-2100F, JEOL Ltd., Tokyo, Japan) was used to analyze the morphology and sizes of the synthesized samples.

### 2.3. Preparation of Standard Solution

Tartrazine powder was dissolved in ultrapure water to prepare the standard stock solution of 100 μg/mL and this was stored at 4 °C. Then, a series of working standard solutions was obtained by a consecutive dilution method with the concentrations ranging from 10 ng/mL to 10 μg/mL.

### 2.4. Pretreatment of Large Yellow Croaker Sample

The large yellow croaker skin was cut into small circular pieces of 0.5 cm in diameter; we then measured the weight of each piece. A total of 200 μL of tartrazine solution with different concentrations were dropped onto the small pieces of skin to cover the surface. Next, stewing was performed. Subsequently, the large yellow croaker skin was transferred to the centrifuge tubes and 4 mL of ultrapure water was added. Then, the mixture was sonicated for 5 min.

To optimize sample preparation and reduce time and cost, we compared three methods in further experiments. Extractions were filtered through a 0.45 μm PTFE microporous film and then the filtrates underwent SERS analysis. This method was performed only with water (W) extraction and the samples directly underwent a type of filtration purification called the W method. For the second method, 1 mL of methanol (M) was added to the deposit protein. After the ultrasound step, the vortexed solution was filtered with a PTFE microfilter; this was called the W+M method. The third method, called the W+M+C method, had the same steps as that of W+M method, except that a further purification of centrifugation (C) at 5000 rpm for 6 min was performed before filtration.

### 2.5. Raman and SERS Measurements

The normal Raman spectrum of tartrazine and the SERS spectra of standard tartrazine solution or extracts were collected using a Nicolet DXR microscopy Raman spectrometer (Thermo Fisher Scientific Inc., Waltham, MA, USA) equipped with a 633 nm He–Ne laser source. A 20× objective at 5 mW laser power was used throughout the experiments.

For the normal Raman spectrum of tartrazine, a small amount of tartrazine was deposited onto a glass slide and squeezed to a thin film before its normal Raman spectrum was recorded.

For SERS spectra collection, samples were prepared by mixing 20 μL of Ag nanowires colloidal solution, 80 μL of standard tartrazine solution or large yellow croaker skin extract and 10 μL of $MgSO_4$ solution (150 mM). The selection of the volume ratio between the Ag nanowires substrate and tartrazine solution, as well as the types (sulfate, chloride, perchlorate) and amount of aggregation agent, were based on our preliminary experiments. Then, 5 μL of aliquot was allowed to evaporate under ambient conditions for SERS detection. The final spectrum was the average of 10 spectra collected from 10 different random positions. This process was repeated in quadruplicate for reproducibility.

### 2.6. Data Analysis

Delight, version 3.2.1, (DSquared Development Inc., LaGrande, OR, USA) was used for the analysis of spectral data. For quantitative analysis, the partial least square (PLS) model was established to predict the analyte concentrations.

The coefficient of determination ($R^2$), the root-mean-square error (RMSE) and the ratio of performance to deviation (RPD) were used to evaluate the model. The higher the $R^2$ value (the closer to 1) or the lower the RMSE value, the better predictability the model has [24].

## 3. Results

### 3.1. Spectral Features of Tartrazine Dye

Figure 1 shows the Raman spectrum of tartrazine solid. The peaks at 485, 618, 638, 1129, 1358, 1477, 1503 and 1599 $cm^{-1}$ were easily recognizable. Among these peaks, the most prominent tartrazine peak at 1599 $cm^{-1}$ was assigned to the in-plane bending of OH, the asymmetric stretching of $COO^-$, the overlapping effect of symmetric stretching and the out-of-plane C–H deformation of two phenyl rings. The other major peak at 1358 $cm^{-1}$ was attributed to the rocking of the phenyl ring, the symmetric stretching of $COO^-$ and the stretching vibrations of –C–N=N–C–. Table 1 shows the assignments of their vibrational bands [25].

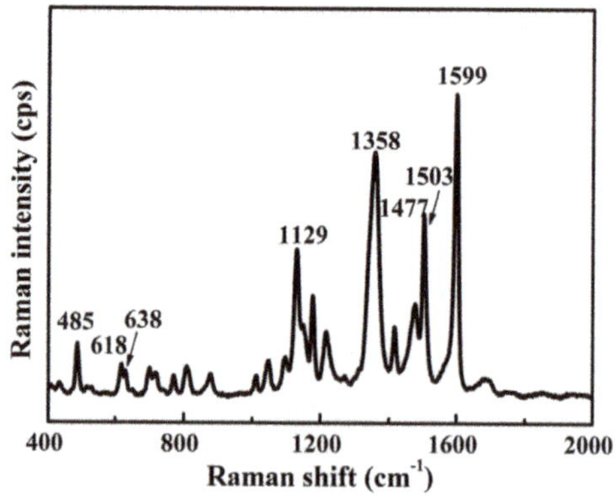


**Figure 1.** Raman spectrum of tartrazine.

**Table 1.** Major band assignment of Raman spectra for tartrazine.

| Raman Shift ($cm^{-1}$) | Vibrational Assignment |
|---|---|
| 485 | $\rho$ ($SO_3^-$), $Def_{op}$ (C–H)$_{ph1, pyr}$ |
| 618 | $Def_{op}$ (C–H)$_{ph1, ph2}$, $\delta$ (pyr) |
| 638 | $Def_{op}$ (ph1, ph2) |
| 1129 | $Def_{op}$ (C–H)$_{ph1, ph2}$, $\delta$ (ph1) |
| 1358 | $\upsilon$ (–C–N=N–C–), $\upsilon_s$ ($COO^-$), $\rho$ (ph2) |
| 1477 | $\delta$ (C–H)$_{ph1, ph2}$, $\upsilon_{as}$ (N=N–C) |
| 1503 | $\delta$ (ph2), $\delta$ (C=C) |
| 1599 | $\upsilon$ (ph1, ph2), $Def_{op}$ (C–H)$_{ph1, ph2}$, $\delta$ (OH), $\upsilon_{as}$ ($COO^-$) |

$\upsilon_s$—symmetric stretching; $\upsilon_{as}$—asymmetric stretching; $\delta$—in-plane bending; $\gamma$—out-of-plane bending; ip—in-plane; op—out-of-plane; $\rho$—rocking; ph—phenyl ring; pyr—pyrazole ring; Def—deformation.

### 3.2. *Ag Nanowire Characterization*

The UV-Vis absorbance spectrum of the as-prepared Ag nanowires exhibits two characteristic absorption peaks at 349 nm and 380 nm, as shown in Figure 2a. The peak located at 349 nm is ascribed to the out-of-plane quadrupole resonance of Ag nanowires while another peak with high intensity at 380 nm is attributed to the out-of-plane dipole resonance of Ag nanowires [26].

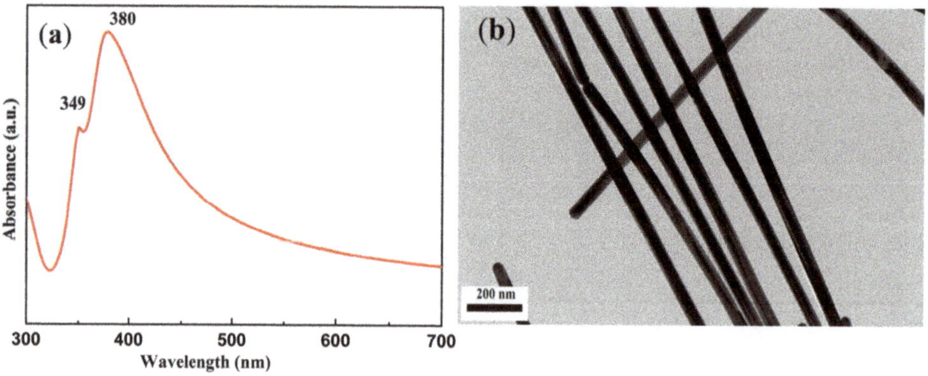


**Figure 2.** (**a**) Ultraviolet-visible (UV-Vis) absorbance spectrum of Ag nanowires; (**b**) transmission electron microscopy (TEM) images of Ag nanowires.

To determine the normal distribution and a certain amount of homogeneity, we carried out TEM tests. The TEM images of Ag nanowires are shown in Figure 2b. The figure shows that the diameter of the as-prepared Ag nanowires was relatively uniform. An average diameter of 49.4 ± 3.9 nm and length of 7–10 μm were observed by calculating 100 nanowires.

### 3.3. SERS Analysis of Tartrazine

Figure 3 shows the SERS spectral features of the standard tartrazine solutions. The characteristic peaks of standard tartrazine solution can be clearly distinguished at a concentration as low as 0.01 μg/mL. The limit of detection, calculated based on HPLC, was 5.2 ng/mL [27], which was close to the minimum concentration visually observed by our SERS method.

The strongest peak of tartrazine was still visible at 1599 $cm^{-1}$. However, some peaks may red shift or blue shift compared with the normal Raman spectrum shown in Figure 1. For instance, the characteristic peak at 1503 $cm^{-1}$ in the Raman spectrum of tartrazine red shifted to 1504 $cm^{-1}$ in the SERS spectra. The characteristic peak at 1477 $cm^{-1}$ in the Raman spectrum of tartrazine red shifted to 1474 $cm^{-1}$ in the SERS spectra. The shifts in the characteristic peaks were due to changes in site and adsorption orientations or to changes in polarizability after interactions between the analyte and substrate [28,29]. The as-prepared Ag nanowires as SERS substrate could greatly enhance Raman scattering signals of tartrazine. Based on the method described by Ru et al. [30], the enhancement factors for tartrazine were calculated as $3.9 \times 10^6$ (based upon the strongest peak at 1599 $cm^{-1}$), the reproducibility of the as-prepared Ag nanowires was reported in our previous study [23].

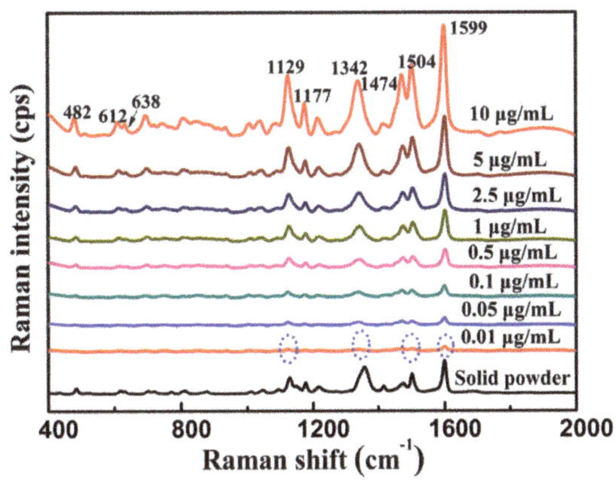


**Figure 3.** Surface-enhanced Raman scattering (SERS) optical spectra of the standard tartrazine solution ($n$ = 40).

As shown in Figure 3, the intensity of the characteristic peaks increased remarkably with increasing concentration. Therefore, the quantitative analysis model of the standard tartrazine solutions may be established based on the SERS spectra. The SERS spectra of tartrazine solutions at eight different concentration levels (10 ng/mL–10 μg/mL) were collected to establish the PLS model ($n$ = 32, four spectra for each concentration). As shown in Figure 4, a high linear correlation ($R^2$ = 0.970) was observed between the actual and the predicted concentrations. High RPD (6.10) and low RMSE (0.26) values also demonstrate that this model has relatively good predictability [16].

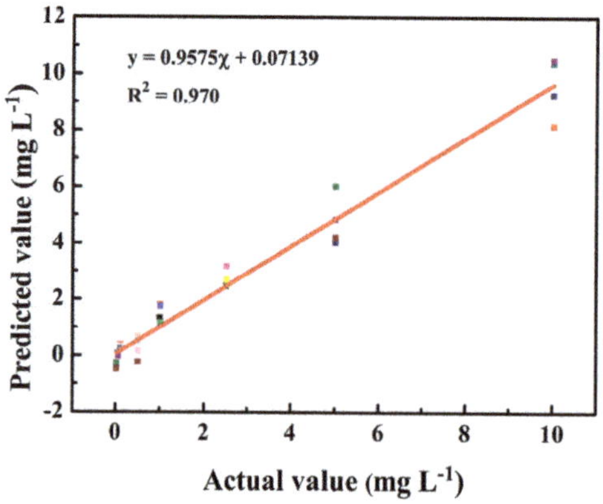


**Figure 4.** Partial least square (PLS) model of the standard tartrazine solution.

### *3.4. SERS Analysis of Tartrazine in Large Yellow Croaker*

To simulate the process of the illegal addition of tartrazine to the large yellow croaker, we experimentally spiked large yellow croaker peel extracts with tartrazine (Figure 5a). We aimed to analyze the large yellow croaker matrix from the tartrazine extracts in the large yellow croaker peel experiment. Figure 5a,b show more impure peaks, such as the peak at 1428 $cm^{-1}$, than in Figure 4 due to the interference of non-targeted components. For the first preparation of the W sample, an approximate 12% decrease was observed at the intensity peak of 1599 $cm^{-1}$ in Figure 5a compared with that in Figure 5b, thereby indicating that the recovery rate of the large yellow croaker was approximately 88% of the sample preparation. Similarly, the recovery rates of the preparation of the W+M and W+M+C samples were 84% and 81%, respectively. From W to W+M+C, the decrease in recovery rate may have been due to the increase in purification steps. As shown in Figure 5b, W+M+C had a significant increase in SERS signal, although it had more purification steps than the afore mentioned methods. The W+M+C method took less than 20 mins to prepare the sample, the decrease in recovery rate was minimal and the method showed positive SERS performance. Therefore, this sample preparation method was chosen as the optimal method for further tartrazine detection in large yellow croaker.

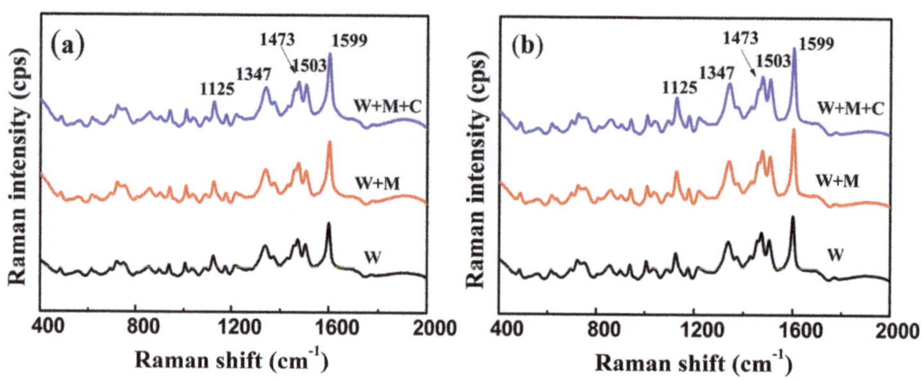


**Figure 5.** *Cont.*

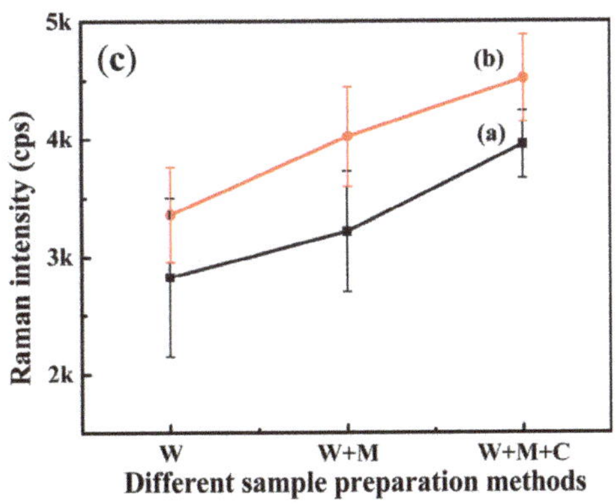


**Figure 5.** (**a**) Surface-enhanced Raman scattering (SERS) spectra extract from large yellow croaker peels spiked with 509.55 ng/cm$^2$ of tartrazine; (**b**) extract added with tartrazine equivalent (a) but assumed 100% recovery rate; (**c**) Raman intensity of tartrazine at 1599 cm$^{-1}$ via different sample preparation methods ($n$ = 20).

Tartrazine is used illegally to color fish to make low-priced fish, such as albiflora croaker, appear similar to high-priced yellow croaker or putrescent yellow croaker. Dripping standard tartrazine solutions on a white paper showed that 2.5 ppm was the lowest critical concentration identified by the naked eye, converting 25.48 ng/cm$^2$ to 200 µL in an area of 19.62 cm$^2$. Hence, the color was completely invisibility when this concentration was added to yellow croaker peels.

We detected tartrazine on the basis of the optimal sample preparation. Figure 6 illustrates the SERS spectra obtained from seven different tartrazine concentrations adsorbed in the large yellow croaker. The spectra revealed that the lowest detectable concentration was 20.38 ng/cm$^2$, which was lower than 25.48 ng/cm$^2$. Thus, the as-prepared Ag nanowires coupled with SERS technology can meet the requirement of zero-tolerance residue limit. Compared to the HPLC method, this method has the advantage of a short detection time. However, there is still a long way to go before the SERS technique overcomes the matrix interference absolutely.

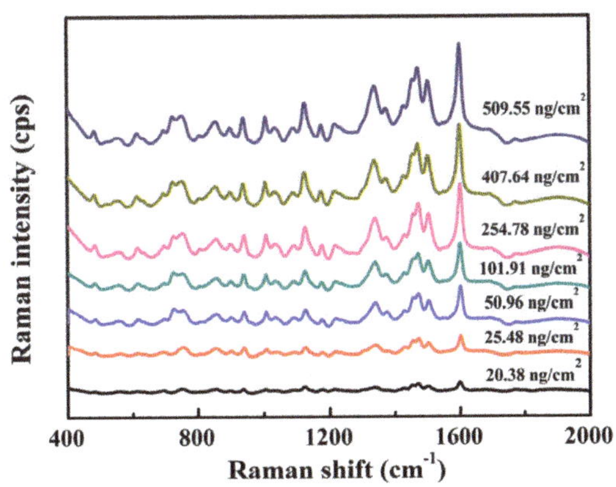


**Figure 6.** Representative surface-enhanced Raman scattering (SERS) spectra of tartrazine extracts from large yellow croaker skin ($n$ = 30).

## 4. Conclusions

In this study, the as-prepared Ag nanowire SERS substrate was tested as a potential tool to determine tartrazine in large yellow croaker for the first time. Highly uniform and high-quality

Ag nanowires were synthesized in a time-efficient and cost-effective manner, which displayed the substantially enhanced effect on standard tartrazine solution detection. The results showed that the lowest detectable concentration was 10 ng/mL. The PLS model demonstrated good predictability. In terms of recovery rate, time, cost and SERS signal, the optimal sample preparation method (i.e., W+M+C) was selected and could be detected at concentration levels as low as 25.48 ng/cm$^2$, which fully satisfies the actual detection requirements. Hence, the established SERS approach showed remarkable performance in determining tartrazine residue in large yellow croaker.

**Author Contributions:** J.S. performed the experiments and wrote the original manuscript. Y.Z. gave theoretical and practical guidance. Y.H. provided worthy advice on work design and modified the manuscript. Y.F. helped design figures and analyzed experimental data. K.L. contributed to the research design, data analysis and manuscript preparation. All authors commented on the manuscript.

**Funding:** This research was funded by the National Key Research and Development Program of China, grant number 2017YFC1600706.

**Conflicts of Interest:** The authors declare no conflicts of interest.

## References

1. Llamas, N.E.; Garrido, M.; Di Nezio, M.S.; Fernandez Band, B.S. Second order advantage in the determination of amaranth, sunset yellow FCF and tartrazine by UV-vis and multivariate curve resolution-alternating least squares. *Anal. Chim. Acta* **2009**, *655*, 38–42. [CrossRef] [PubMed]
2. Gupta, V.K.; Jain, R.; Nayak, A.; Agarwal, S.; Shrivastava, M. Removal of the hazardous dye—Tartrazine by photodegradation on titanium dioxide surface. *Mater. Sci. Eng. C* **2011**, *31*, 1062–1067. [CrossRef]
3. Rowe, K.S.; Rowe, K.J. Synthetic food coloring and behavior: A dose response effect in a double-blind, placebo-controlled, repeated-measures study. *J. Pediatr.* **1994**, *125*, 691–698. [CrossRef]
4. Guyot, S.; Serrand, S.; Le Quéré, J.M.; Sanoner, P.; Renard, C.M.G.C. Enzymatic synthesis and physicochemical characterisation of phloridzin oxidation products (POP), a new water-soluble yellow dye deriving from apple. *Innov. Food Sci. Emerg.* **2007**, *8*, 443–450. [CrossRef]
5. National Food Safety Standard: Standards for the Uses of Food Additives (GB2760-2014). Available online: http://www.nhfpc.gov.cn/zwgkzt/cybz/201412/d9a9f04bc35f42ecac0600e0360f8c89.shtml (accessed on 7 August 2018).
6. A List of Non-Edible Substances that May Be Added Illegally and Food Additives that May Be Abused (Summary of Batches 1–5). Available online: http://www.nhfpc.gov.cn/sps/s7892/201406/38e5c8a53615486888d93ed05ac9731a.shtml (accessed on 7 August 2018).
7. Elena, D.; Cornelia, P.E. Study of analytical parameters of the HPLC method for tartrazine and sunset yellow analysis in soft drinks. *Rev. Chim. Buchar.* **2010**, *61*, 1177–1182.
8. Chen, X.; Zhao, Y.; Shen, H.; Zhou, L.; Pan, S.; Jin, M. Fast determination of seven synthetic pigments from wine and soft drinks using magnetic dispersive solid-phase extraction followed by liquid chromatography-tandem mass spectrometry. *J. Chromatogr. A* **2014**, *1346*, 123–128. [CrossRef] [PubMed]
9. Saleh, M.M.S.; Hashem, E.Y.; Al-Salahi, N.O.A. Oxidation and complexation-based spectrophotometric methods for sensitive determination of tartrazine E102 in some commercial food samples. *Comput. Chem.* **2016**, *4*, 51–64. [CrossRef]
10. Gomez, M.; Arancibia, V.; Rojas, C.; Nagles, E. Adsorptive stripping voltammetric determination of tartrazine and sunset yellow in gelatins and soft drink powder in the presence of cetylpyridinium bromide. *Int. J. Electrochem. Sci.* **2012**, *7*, 7493–7502.
11. De Andrade, F.I.; Guedes, M.I.F.; Vieira, Í.G.P.; Mendes, F.N.P.; Rodrigues, P.A.S.; Maia, C.S.C.; Ávila, M.M.M.; Ribeiro, L.d.M. Determination of synthetic food dyes in commercial soft drinks by TLC and ion-pair HPLC. *Food Chem.* **2014**, *157*, 193–198. [CrossRef] [PubMed]
12. Ou, Y.; Pei, L.; Lai, K.; Huang, Y.; Rasco, B.A.; Wang, X.; Fan, Y. Rapid analysis of multiple Sudan dyes in chili flakes using surface-enhanced Raman spectroscopy coupled with Au–Ag core-shell nanospheres. *Food Anal. Methods* **2017**, *10*, 565–574. [CrossRef]
13. Xu, M.; Gao, Y.; Han, X.X.; Zhao, B. Detection of pesticide residues in food using surface-enhanced Raman spectroscopy: A review. *J. Agric. Food Chem.* **2017**, *65*, 6719–6726. [CrossRef] [PubMed]

14. Fan, Y.; Lai, K.; Rasco, B.A.; Huang, Y. Analyses of phosmet residues in apples with surface-enhanced Raman spectroscopy. *Food Control* **2014**, *37*, 153–157. [CrossRef]
15. Liu, B.; Zhou, P.; Liu, X.; Sun, X.; Li, H.; Lin, M. Detection of pesticides in fruits by surface-enhanced Raman spectroscopy coupled with gold nanostructures. *Food Bioprocess Technol.* **2013**, *6*, 710–718. [CrossRef]
16. Luo, H.; Huang, Y.; Lai, K.; Rasco, B.A.; Fan, Y. Surface-enhanced Raman spectroscopy coupled with gold nanoparticles for rapid detection of phosmet and thiabendazole residues in apples. *Food Control* **2016**, *68*, 229–235. [CrossRef]
17. Futamata, M.; Yu, Y.; Yajima, T. Elucidation of electrostatic interaction between cationic dyes and Ag nanoparticles generating enormous SERS enhancement in aqueous solution. *J. Phys. Chem. C* **2011**, *115*, 5271–5279. [CrossRef]
18. Zhang, Y.; Huang, Y.; Zhai, F.; Du, R.; Liu, Y.; Lai, K. Analyses of enrofloxacin, furazolidone and malachite green in fish products with surface-enhanced Raman spectroscopy. *Food Chem.* **2012**, *135*, 845–850. [CrossRef] [PubMed]
19. Ai, Y.; Liang, P.; Wu, Y.; Dong, Q.; Li, J.; Bai, Y.; Xu, B.; Yu, Z.; Ni, D. Rapid qualitative and quantitative determination of food colorants by both Raman spectra and surface-enhanced Raman scattering (SERS). *Food Chem.* **2018**, *241*, 427–433. [CrossRef] [PubMed]
20. Saviello, D.; Di Gioia, A.; Turenne, P.-I.; Trabace, M.; Giorgi, R.; Mirabile, A.; Baglioni, P.; Iacopino, D. Handheld su31rface-enhanced Raman scattering identification of dye chemical composition in felt-tip pen drawings. *J. Raman Spectrosc.* **2018**. [CrossRef]
21. Saviello, D.; Trabace, M.; Alyami, A.; Mirabile, A.; Giorgi, R.; Baglioni, P.; Iacopino, D. A combined surface enhanced Raman spectroscopy (SERS)/UV-vis approach for the investigation of dye content in commercial felt tip pens inks. *Talanta* **2018**, *181*, 448–453. [CrossRef] [PubMed]
22. Meng, J.; Qin, S.; Zhang, L.; Yang, L. Designing of a novel gold nanodumbbells SERS substrate for detection of prohibited colorants in drinks. *Appl. Surf. Sci.* **2016**, *366*, 181–186. [CrossRef]
23. Song, J.; Huang, Y.; Fan, Y.; Zhao, Z.; Yu, W.; Rasco, B.; Lai, K. Detection of prohibited fish drugs using silver nanowires as substrate for surface-enhanced Raman scattering. *Nanomaterials* **2016**, *6*, 175. [CrossRef] [PubMed]
24. Huang, Y.; Cavinato, A.G.; Mayes, D.M.; Bledsoe, G.E.; Rasco, B.A. Nondestructive prediction of moisture and sodium chloride in cold smoked Atlantic salmon (*Salmo salar*). *J. Food Sci.* **2002**, *67*, 2543–2547. [CrossRef]
25. Peica, N.; Pavel, I.; Pînzaru, S.C.; Rastogi, V.K.; Kiefer, W. Vibrational characterization of E102 food additive by Raman and surface-enhanced Raman spectroscopy and theoretical studies. *J. Raman Spectrosc.* **2005**, *36*, 657–666. [CrossRef]
26. Gao, Y.; Song, L.; Jiang, P.; Liu, L.F.; Yan, X.Q.; Zhou, Z.P.; Liu, D.F.; Wang, J.X.; Yuan, H.J.; Zhang, Z.X.; et al. Silver nanowires with five-fold symmetric cross-section. *J. Cryst. Growth* **2005**, *276*, 606–612. [CrossRef]
27. Sha, O.; Zhu, X.; Feng, Y.; Ma, W. Determination of sunset yellow and tartrazine in food samples by combining ionic liquid-based aqueous two-phase system with high performance liquid chromatography. *J. Anal. Methods Chem.* **2014**, *2014*, 964273. [CrossRef] [PubMed]
28. Seney, C.S.; Gutzman, B.M.; Goddard, R.H. Correlation of size and surface-enhanced Raman scattering activity of optical and spectroscopic properties for silver nanoparticles. *J. Phys. Chem. C* **2009**, *113*, 74–80. [CrossRef]
29. Nardou, E.; Vouagner, D.; Jurdyc, A.M.; Berthelot, A.; Pillonnet, A.; Sablonière, V.; Bessueille, F.; Champagnon, B. Surface enhanced Raman scattering in an amorphous matrix for Raman amplification. *J. Non-Cryst. Solids* **2011**, *357*, 1895–1899. [CrossRef]
30. Ru, E.C.L.; Blackie, E.; Meyer, M.; Etchegoint, P.G. Surface enhanced Raman scattering enhancement factors: A comprehensive study. *J. Phys. Chem. C* **2007**, *111*, 13794–13803.

nanomaterials

MDPI

Article

# Application of Self-Assembled Raman Spectrum-Enhanced Substrate in Detection of Dissolved Furfural in Insulating Oil

**Haiyang Shi [1,*], Weigen Chen [1], Fu Wan [1,*], Lingling Du [2], Shuhua Zhang [1], Weiran Zhou [1], Jiayi Zhang [1], Yingzhou Huang [3] and Chengzhi Zhu [4]**

1. State Key Laboratory of Power Transmission Equipment & System Security and New Technology, Chongqing University, Chongqing 400044, China; weigench@cqu.edu.cn (W.C.); zhangshuhua@cqu.edu.cn (S.Z.); zhouweiran@cqu.edu.cn (W.Z.); vinisbaby@163.com (J.Z.)
2. Chengdu Power Supply Company, Chengdu 610041, China; dulingling2014@163.com
3. Soft Matter and Interdisciplinary Research Center, Chongqing University, Chongqing 400044, China; yzhuang@cqu.edu.cn
4. State Grid Zhejiang Electric Power Company, Hangzhou 310000, China; yth11322@foxmail.com

* Correspondence: shihaiyang@cqu.edu.cn (H.S.); fuwan@cqu.edu.cn (F.W.); Tel.: +86-651-117-95 (H.S. & F.W.); Fax: +86-651-024-42 (H.S. & F.W.)

Received: 21 November 2018; Accepted: 19 December 2018; Published: 23 December 2018

**Abstract:** Accurate detection of dissolved aging features in transformer oil is the key to judging the aging degree of oil-paper insulation. In this work, in order to realize in situ detection of furfural dissolved in transformer oil, silver nanoparticles were self-assembled on the surface of gold film with P-aminophenylthiophenol (PATP) as a coupling agent. Rhodamine-6G (R6G) was used as the probe molecule to test the enhancement effect. By optimizing the molecular concentration, molecular deposition time, and silver sol deposition time of PATP, the nanoparticles were made more uniform and compact, and an enhanced substrate with rich hot spots was obtained. The optimum substrate was developed, and surface-enhanced Raman spectroscopy (SERS) detection of trace furfural dissolved in transformer oil was realized. The results showed that the substrate prepared under the conditions of 0.1 mol/L PATP, 5 h deposition in PATP and 12 h immersion in silver sol, had the best reinforcement effect (that is, uniform and compact particle arrangement and no particle clusters). By use of this substrate, the minimum detectable concentration of furfural in transformer oil was about 1.06 mg/L, which provides a new method for fast and nondestructive detection of transformer aging diagnosis.

**Keywords:** surface-enhanced Raman spectroscopy; transformer aging; concentration detection

## 1. Introduction

Power transformers are indispensable equipment in a power system. Transformer failures occurring during service cause heavy economic losses and serious casualties to power suppliers. Therefore, regular diagnosis of transformer health is of great importance, especially for aging transformers [1–3]. The effects of temperature, electric field, water, and oxygen on the long-term operation of a transformer results in the breakdown of cellulose chains in insulating paper and produces furan derivatives dissolved in transformer oil [4–7]. It is generally believed that glucose monomers that break the ends of cellulose chains are unstable and easy to break away from cellulose chains in the process of cellulose degradation. The disintegrated glucose monomers are easily decomposed by heating, causing five furan compounds including furfural, acetyl-furan, methyl furfural, furaldehyde, and 2,5-hydroxyl-methyl-furan formaldehyde. Among them, the content of furfural is the highest [8–10]. Thus, it is one of the most commonly used indexes to evaluate the insulation aging of oil paper. The aging of insulating oil does not produce furfural, which is produced only by the breakage of cellulose chains in the aging process of insulating paper. Furfural in oil is determined by insulating paper. Therefore, furfural in oil exclusively reflects the aging degree of insulating paper [11–14]. This is a great advantage (furfural analysis in oil) in evaluating the aging of insulating paper. In addition, some study results have shown that furfural production is directly related to the aging of insulating paper under normal or overheated conditions compared to four other furan compounds, so furfural content could target the characteristics of the aging of insulating paper [15–18]. At the same time, another advantage of furfural detection is easy to measure. According to field transformer data, the content of furfural in oil is usually much higher than four other furan compounds, which makes it easy to accurately detect. In 1996, the content of furfural was considered to be one of the necessary testing items in document DL/T596—1996. It pointed out that the aging level of a transformer is in the middle stage of life when the concentration of furfural dissolved in oil reaches 0.5 mg/L, and the aging level of the transformer is in late life when the concentration of dissolved furfural in the oil reaches 4 mg/L [19]. Therefore, the analysis of the dissolved furfural content in transformer oil is of great significance to the correct evaluation of the aging state of a transformer. At present, the detection methods of furfural content in oil mainly include spectrophotometry and high-performance liquid chromatography [20–22]. These detection techniques have high detection accuracy, but they all need to extract transformer oil samples and other pretreatment. They require high requirements for operators and the detection environment, complex operation processes, and long detection cycles, which can only be completed in laboratory operations. A new detection method is an urgent need for in-site and rapid detection of dissolved furfural concentration in transformer oil. Raman spectroscopy is a spectroscopic method used to detect the vibration of molecules. The main principle of Raman spectroscopy is inelastic scattering of light irradiated on matter. In inelastic collisions, energy exchange occurs between photons and molecules. Photons not only change the direction of motion, but also transfer part of the energy to molecules. The vibrational energy of a molecule is transmitted to the photon, changing the frequency of the photon. This scattering process is called Raman scattering. Raman spectroscopy, as a single wavelength laser detection technology, has the advantages of no sample pretreatment, no loss of samples, and fast detection speed [23–27]. In 2015, Somekawa et al. measured the content of dissolved furfural in oil by laser Raman spectroscopy, and realized the measurement of furfural with a minimum detection concentration of 14.4 mg/L in oil [28]. In 2016, Gu et al. realized the detection of furfural in 0.1 mg/L by using confocal Raman technology and extraction technology, and achieved a maximum detection error of not more than 12.04% [29]. However, it is difficult to detect the low concentration of small molecules due to smaller molecular cross-sections. Fleischmann et al. found that pyridine molecules adsorbed on roughened Ag electrodes exhibited a large Raman scattering phenomenon [30]. In addition, the selective adsorption molecules on the active carrier surface inhibit the fluorescence emission, which greatly improves the signal-to-noise ratio of laser Raman spectroscopy. This surface enhancement effect is called surface-enhanced Raman scattering (SERS). The SERS mechanism mainly includes electromagnetic field enhancement and chemical enhancement, in which electromagnetic field enhancement is dominant. This enhancement is produced by a surface plasmon resonance effect (that is, free electrons in metals have a collective oscillation effect under the action of optical and electrical fields [31–34]). In recent years, SERS has been widely used in surface adsorption, electrochemical and catalytic reactions, chemical and biological sensors, biomedical detection, trace detection, and substance analysis [35–42]. In addition, a large number of new methods for fabricating enhanced substrates have been studied. Sergio et al. prepared uniform gold nano-octahedron structures combined with the use of a microfluidic technique based on micro-evaporation [43]. Jeong et al. prepared silver nanoshells with magnetic and SERS properties, which have been used to detect trace amounts of organic molecules [44]. Therefore, it is pretty meaningful to apply SERS technology to the detection of furfural in transformer oil.

In this paper, the effects of P-aminophenylthiophenol (PATP) concentration, deposition time in PATP, and immersion time in silver sol on the reinforcing properties of substrates were studied. The surface morphology of the substrate was characterized by X-ray photoelectron spectroscopy (XPS) and scanning electron microscopy (SEM). Rhodamine-6G (R6G) was used as the probe molecule to test the enhancement effect. By optimizing the PATP molecular concentration, PATP molecular deposition time, and silver sol deposition time, the nanoparticles were more uniform and compact, and an enhanced substrate with rich hot spots was obtained. The low concentration and in situ detection of dissolved furfural in transformer oil were realized, which provides a new method for fast and nondestructive detection of transformer aging diagnosis.

## 2. Experimental Part

Silver nitrate ($AgNO_3$), P-aminophenol (PATP), sodium citrate, and Rhodamine-6G (R6G) were purchased from Aladdin (Shanghai, China). Furfural and transformer oil (Karamay 25#) were purchased from Chuan Dong chemical company of Chongqing, China. Transformer oil was used as a solvent to prepare the sample solution. In order to avoid the influence of the initial state of new oil on the experimental results, degassing and drying of the new oil were carried out before experimentation. Twenty-five milligrams of furfural were fully dissolved in 225 mL of transformer oil, and 100 mg/L standard sample solution was obtained. In order to prevent the change of mass fraction caused by the decomposition of furfural under visible light, the prepared solution was quickly sealed in a brown reagent bottle and stored in the dark. Furfural transformer oil solution of 100 mg/L could be diluted with new transformer oil proportionately to obtain furfural samples with different concentrations dissolved in the oil. The standard sample solution was diluted proportionately to obtain different concentration samples. Using the magnetron sputtering technique, a gold shell with a thickness of 100 nm was deposited on the silicon chip. Silver nanoparticles were synthesized using the sodium citrate reduction method according to Lee-Meisel [45]. The gold film was washed in water and ethanol solution for 20 min each, then immersed in PATP solution for a period of time, removed with tweezers, and washed repeatedly to remove the surface of the unbounded PATP molecules. The pretreated gold film was dried with nitrogen and soaked in silver sol for a certain time, then washed several times alternately with ethanol and deionized water and dried with nitrogen at room temperature for storage. All of the SERS experiments in this work were measured by a commercial Micro-Raman spectrometer (ANDOR, SR-5000i-C, Oxford, England), and 532 nm lasers were chosen to be the illuminating sources. The integral time of the spectrometer was set at 10 s, and the accumulated integral was 3 times. In addition, the 1200 L/mm grating and the 100-m slit width of the spectrometer were used to detect the sample. The morphology and sizes of the silver nanostructures were characterized by scanning electron microscopy (SEM, TESCAN Mira3 LMH, Brno, Czech Republic). X-ray photoelectron spectroscopy (XPS, Thermo-Fisher-Scientific ESCALAB250Xi, Shanghai, China) was used to analyze the chemical valence states of elements on the surfaces of the substrates.

## 3. Results and Discussion

Using a magnetron sputtering technique, a gold shell with a thickness of 100 nm was deposited on a silicon chip, and then PATP molecules with special functional groups were modified on the metal membrane. Finally, the prepared silver nanoparticles were deposited on the surface in order to obtain the SERS-enhanced substrate of the sandwich structure. A schematic diagram of the specific process is shown in Figure 1a. The main reason for choosing gold film as a substrate was to format a dense PATP molecular membrane on the surface. Sulfhydryl compounds and gold films can be well bonded to form S-Ag bonds. Silver nanoparticles rely on N-Ag bonds formed by adsorbed PATP functional groups, which can avoid an agglomeration effect. It was necessary to observe the deposition of silver nanoparticles on the surface of the gold film by scanning electron microscopy, considering the uneven distribution of common deposition methods. In Figure 1b, the gold film was basically covered by silver nanoparticles, which was very important for testing data stability. Due to the gold film being covered with silver nanoparticles, this indicated that adsorption of PATP molecules effectively avoided the aggregation of silver nanoparticles. It was beneficial to enhance the uniformity of the distribution of hot spots, which was particularly an important point for later experimental tests. In addition, the size and the shape of the particles had a great influence on the enhancement of the substrate. The preparation of silver nanoparticles was susceptible to the influence of temperature and external environment, which needed to be avoided as much as possible. By calculating the particle size distribution of silver nanoparticles in electron micrographs by using software (Nano Measure), it was found that the particle size was mainly concentrated in the range of 50–60 nm (Figure 1e). More importantly, in order to further determine the adsorption mode of PATP molecules on the surface of gold film, X-ray photoelectron spectroscopy tests on the modified gold film (Figure 1c, red line) were carried out. Compared to the unmodified gold film (Figure 1c, black line ($Au4d_{5/2}$ and $Au4d_{3/2}$ represent the presence of gold)), the characteristic lines of elemental carbons (C1s) and nitrogens (N1s) appeared, which indicated that the PATP molecules had been effectively adsorbed on the surface of the gold films. Figure 1d shows the N1s' narrow spectrum of modified gold film. It is obvious that the characteristic peaks only appeared at 398.7 ev, which indicates that the amino groups on the PATP molecule did not react with the gold film. The reason was that new characteristic peaks appeared at 400.7 ev when the dehydrogenation of amino group occurred. This confirmed that the sulfhydryl group on the PATP molecule formed an Au-S bond with the gold film, while the amino group was far away from the gold film.

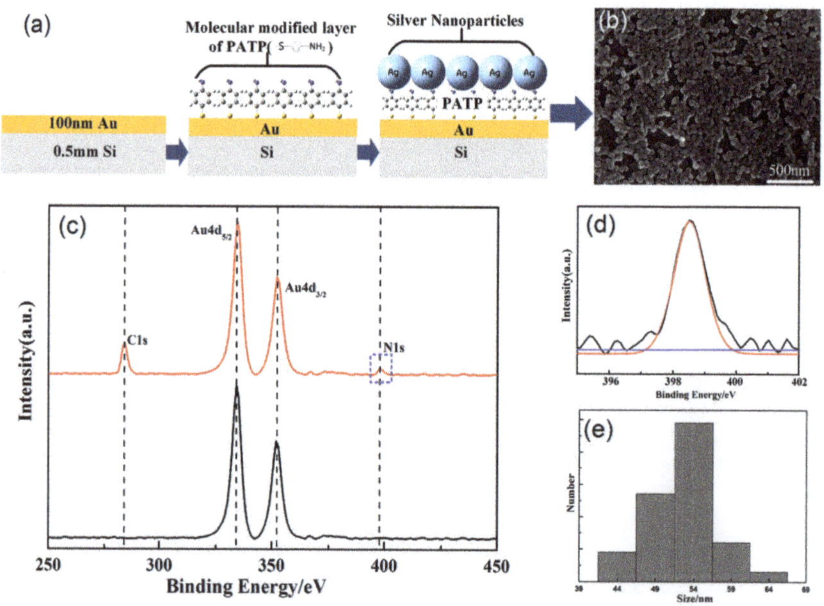


**Figure 1.** (**a**) Schematic diagram of enhanced substrate preparation; (**b**) SEM image of a surface-enhanced Raman scattering (SERS) substrate; (**c**) X-ray photoelectron spectroscopy (XPS) tests on modified gold film (red line) and no modified gold film (black line); (**d**) XPS nitrogen (N1s) spectrum for modified gold film; (**e**) size distribution of the Ag nanoparticles.

The enhancement effect mainly came from the surface plasmon resonance (SPR) produced by metal nanostructures under laser irradiation, and was mainly related to the metal material, morphology, substrate, particle size, and spacing. Previous studies have shown that the surface enhancement effects basically disappear when nanoparticle spacing is greater than 10 nm. Therefore, the substrate must satisfy the characteristics of high sensitivity and homogeneity, not only to meet the tight arrangement between particles, but also to avoid large area agglomeration effects. In the process of substrate preparation, PATP molecular concentration, molecular deposition time, and silver sol deposition time directly affect the final aggregation state of silver nanoparticles. Therefore, it was essential to further study the optimal experimental conditions. In Figure 2a, the basement and corresponding scanning electron microscope are given at PATP concentrations of 0.03, 0.05, and 0.1 mol/L, respectively. The distribution of silver nanoparticles on the substrate surface was more uniform and compact with increasing concentration. In order to evaluate the enhancement effect of the three substrates, the SERS properties of the substrates were tested for $1 \times 10^{-6}$ mol/L R6G ethanol solution as a probe molecule. The results are shown in Figure 2a. It is seen that the prepared substrate had an obvious enhancement effect on R6G. The SERS intensity of R6G on the substrate prepared at a PATP concentration of 0.1 mol/L was the strongest, which means that the substrate prepared at a PATP concentration of 0.1 mol/L had the highest SERS activity and the best enhancement effect, and these consequences were consistent with the above characterization results. This may have been due to the strong local electromagnetic field generated when the nanoparticles were close enough to form a "hot spot", which greatly enhanced the Raman scattering signal of the probe molecules located at the "hot spot". In addition, the deposition time of the substrate in the solution also affected the experimental results. Figure 2b gives the SEM of five substrates prepared in the deposited solution for 1, 3, 5, 7, and 9 h, respectively. The graphs show the silver nanoparticle coverage on the surface of the gold film increased and became more uniform, from 1 to 5 h. When the substrate was deposited in solution for more than 5 h, the coverage of silver nanoparticles on the surface of the gold film increased further, but the partially agglomerated particles appeared. R6G was used as a probe molecule to test the substrate properties in Figure 2b. The results showed that the substrate enhancement effect was the best when deposited for 5 h. This may have been because with the increase of deposition time in PATP ethanol solution, the coverage of particles on the substrate surface became higher and higher, and the spacing between particles decreased, forming a large number of "hot spots". However, with the increase in deposition time, the agglomeration of metal nanoparticles occurred, which led to a decrease in the reinforcement effect. The immersion of the substrate in solution was of great significance due to the diffusion, collision, adsorption, and binding process of silver nanoparticles under various forces. Under the action of gravity and thermodynamics, the metal particles diffused and collided, and reacted with the gold film by the adsorption of amino groups near the substrate. Thus, it was particularly important to select the best soaking time. Figure 2d shows SEM images of six substrates prepared in silver sol for 4, 8, 12, 16, 20, and 24 h, respectively. It is seen from the graph that the coverage of particles on the substrate surface under 4 h of deposition time was small, while the spacing of particles was large, so it could have been difficult to form an effective "hot spot". With the increase of the soaking time of the substrate in silver sol, the coverage of particles on the substrate surface increased and the spacing of particles decreased, so the number of "hot spots" increased. However, as the soaking time became longer, the coverage of particles on the surface increased slowly. The main reason was that the surface coverage of particles increased as time went by during the initial adsorption. However, after a period of time, the surface coverage of the particles was almost saturated due to electrostatic repulsion between particles, so it could not be increased if the depositing time was too long. In addition, the results indicated that the enhancement effect of substrate on R6G increased with time added at a range from 4 to 12 h, but the enhancement effect was no longer obvious after 12 h. This may have been because the coverage of particles on the substrate surface was getting higher and higher, and the "hot spot" was getting richer, so the reinforcement effect on R6G was getting better and better. After 12 h of deposition, the coverage of particles on the surface was close to saturation, and the enhancement effect was barely changed. Hence, silver nanoparticles ought to have been uniformly and tightly arranged on the surface-enhanced substrate prepared by immersing in silver sol for 12 h, which had a good reinforcing effect and shortened the experimental period. According to the analysis based on the above experimental results, it could be found that there was a direct relationship between the coverage of silver nanoparticles on the substrate surface and the enhancement effect of the substrate. The aggregation state of silver nanoparticles on the substrate surface could be changed by changing the self-assembly parameters (e.g., PATP concentration, deposition time in PATP, immersion time in silver sol), and different SERS substrates with different enhanced effects could be prepared. The substrate with the best enhancement effect could be prepared by use of this method, which needs no special equipment, is simple to operate, has a low cost, and can be used to control the morphology of the substrate by various means. In subsequent experiments, we prepared the gold film as substrate for surface-enhanced Raman analysis of trace characteristics dissolved in transformer oil, under the conditions of 0.1 mol/L PATP concentration, 5 h deposition in PATP, and 12 h immersion in silver gel (More details of the substrate were included in the Supplementary, Figures S1 and S2).

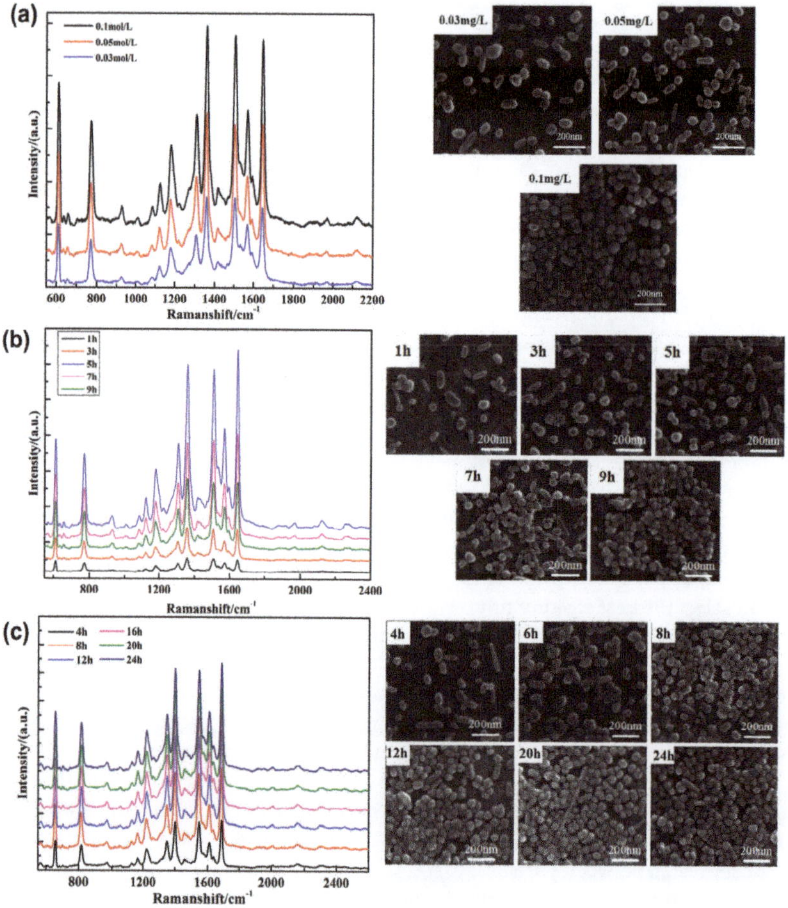


**Figure 2.** SERS spectra of $1 \times 10^{-6}$ mol/L Rhodamine-6G (R6G) with substrates for (**a**) different P-aminophenylthiophenol (PATP) concentrations; (**b**) different deposition times in PATP; and (**c**) different immersion times in silver sol.

As a kind of mineral insulating oil, transformer oil mainly consists of alkanes, cycloalkanes, and unsaturated hydrocarbons, so its Raman spectrum is more complicated and there is some fluorescence interference. As shown in Figure 3a, black lines represent part of the spectrum of transformer oil, and red lines represent the Raman spectra of the 100 mg/L furfural in transformer oil. The comparison results show that the Raman signal of furfural was completely submerged by transformer oil, and the reason was the complex composition of transformer oil. Furthermore, blue lines represent the SERS of

100 mg/L furfural concentration in transformer oil. This shows that the Raman signal of furfural could be enhanced effectively by a substrate. To further determine the Raman signal characteristic peaks of furfural on the substrate, the Raman spectra of pure furfural and substrate are given in Figure 3b. Preliminary comparisons show that the SERS spectra mainly came from the signal of the substrate itself, that is, the Raman peak of the PATP molecule. As shown in Table 1, the characteristic peaks of 1072, 1139, 1183, 1386, 1431, and 1570 $cm^{-1}$ belonged to the characteristic peaks of the coupling molecules of PATP. The reason was that PATP was used to connect silver nanoparticles with gold film, and it was affected by electromagnetic coupling and enhanced the Raman signal. Therefore, PATP molecules appeared as strong Raman signals in the spectrum. At the same time, we found that these Raman signals overlapped severely with the Raman spectrum peaks of furfural, which brought serious interference to furfural detection. The Raman spectra of 1229, 1272, 1570, and 1662 $cm^{-1}$ were from furfural, which corresponded to 1224, 1281, 1569, and 1670 $cm^{-1}$ of pure furfural. This meant that the self-assembled surface-enhanced substrates on the gold film could effectively enhance the Raman signal of dissolved furfural in transformer oil, and the Raman wave number shift appeared in these four Raman spectra peaks. This was because the Raman vibration of furfural was affected by the molecule of transformer oil and the surface-enhanced substrate. In addition, the Raman-shift of these four Raman peaks was different, mainly due to physical enhancement and chemical enhancement, so the enhanced substrate had different effects toward the molecule in various modes of vibration. Among these furfural-enhanced Raman signals, the Raman peaks at 1662 $cm^{-1}$ were enhanced, which may have been due to the fact that furfural molecules adsorbed mainly on the surface of silver nanoparticles through oxygen atoms, and the Raman peaks at 1662$cm^{-1}$ were related to the vibration of oxygen atoms. Therefore, it could be preliminarily determined that furfural existed in transformer oil.

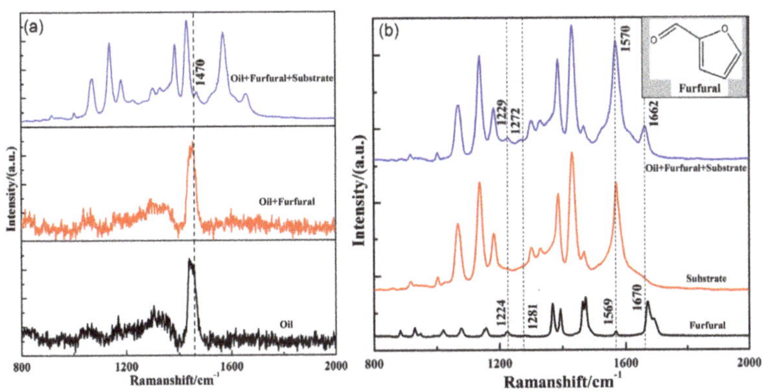


**Figure 3.** (**a**) SERS spectrum of oil with 100 mg/L furfural (Oil + Furfural + Substrate, blue line); Raman spectra of transformer oil with 100 mg/L furfural (Oil + Furfural, red line) and pure transformer oil (black line); (**b**) SERS spectrum of oil with 100 mg/L furfural (Oil + Furfural + Substrate, blue line) and substrate (Substrate, red line); Raman spectrum of pure furfural (Furfural, black line).

It was necessary to know the characteristic Raman peaks of each substance and select the corresponding characteristic Raman peaks for the accurate qualitative and quantitative analysis of each substance. According to the selection principle of Raman characteristic peaks, the characteristic Raman peaks of selected substances should not overlap with the peaks of other components and should have high intensity, and should be within the detection range of a Raman spectrometer. The Raman peaks at 1570 $cm^{-1}$ overlapped with the base self-Raman signals. The Raman peaks at 1229, 1272, and 1662 $cm^{-1}$ were relatively independent, but the intensity of the Raman peaks at 1662 $cm^{-1}$ was higher. In order to obtain better detection sensitivity, 1662 $cm^{-1}$ was selected as the characteristic peak of dissolved furfural molecule in transformer oil for further analysis. There overlapped between 1662 and 1570 $cm^{-1}$, and Lorentzian fitting was used to separate the peaks. The goodness of fit was

0.9984, as shown in Figure 4b. A commonly used method for quantitative analysis of spectral data is the internal standard method. When using the Raman spectroscopy technique to quantitatively analyze the sample solution, the concentration difference of the sample solution, the influence of solvent noise, laser intensity, and other factors interfere with the Raman signal. The absolute peak intensity of the Raman signal fluctuates with the measurement environment, so there is a big error in quantitative analysis this way. While the internal standard substance and the sample are under the same experimental conditions, it can effectively decrease some environmental factors to use the internal standard method for quantitative analysis of the Raman spectra.

**Table 1.** Identification of characteristic peaks in SERS spectra.

| Number | Ramanshift ($cm^{-1}$) | Peak Assignment |
|---|---|---|
| 1 | 919 | Raman signal of substrate and transformer oil |
| 2 | 1004 | Raman signal of substrate and transformer oil |
| 3 | 1072 | Raman signal of substrate |
| 4 | 1139 | Raman signal of substrate |
| 5 | 1183 | Raman signal of substrate |
| 6 | 1229 | Raman signal of Furfural |
| 7 | 1272 | Raman signal of Furfural |
| 8 | 1301 | Raman signal of substrate and transformer oil |
| 9 | 1334 | Raman signal of substrate and transformer oil |
| 10 | 1386 | Raman signal of substrate |
| 11 | 1431 | Raman signal of substrate |
| 12 | 1469 | Raman signal of substrate |
| 13 | 1570 | Raman signal of substrate and Furfural |
| 14 | 1662 | Raman signal of Furfural |

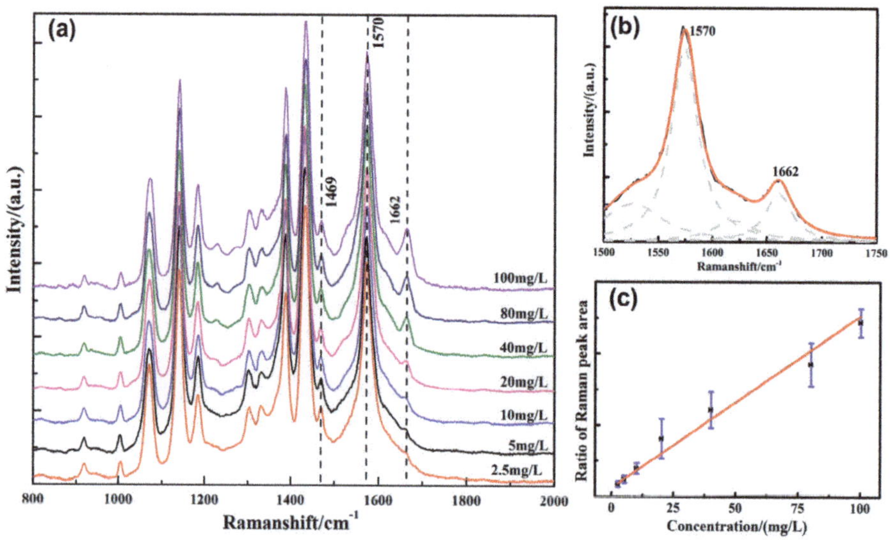


**Figure 4.** (**a**) Raman characteristic peaks of furfural at different concentrations dissolved in transformer oil; (**b**) multipeak fitting of overlapping peaks; (**c**) Raman peak area ratio ($A_{1662\,cm^{-1}}/A_{1469cm^{-1}}$) as a function of dissolved furfural concentration.

In this paper, the internal standard method was used for quantitative analysis. In SERS spectra of the sample solution, the furfural peak at 1662 $cm^{-1}$ was taken as a quantitative peak, and the substrate peak at 1469 $cm^{-1}$ was taken as the internal reference peak. The calibration curve equation was established by the least square fitting method, and then the substrate concentration to be measured was calculated on the basis of the ratio of the Raman peaks area. Selecting the Raman peak of substrate as the internal peak could not only avoid errors in quantitative analysis, but could reduce the interference of adding materials. After removing the baseline with baseline correction, a Savitzky–Golay polynomial was used to smooth out the noise, as shown in Figure 4a. Each Raman peak intensity at 1229, 1272, 1570, and 1662 $cm^{-1}$ reduced with the decrease in furfural concentration. By using the least square method, linear regression was performed on the ratio of the furfural characteristic peak area at 1662 $cm^{-1}$ to the substrate Raman peak area at 1469 $cm^{-1}$ ($A_{1662\ cm^{-1}}/A_{1469\ cm^{-1}}$) and the concentration of dissolved furfural in transformer oil. The results are shown in Figure 4c. It is seen from the graph that there was a good linear relationship between the ratio of the Raman peak area and the concentration of furfural in the range of measured concentration, and the goodness of fit was 0.98. The linear regression equation was:

$$y = 0.44\,x + 0.42. \quad (1)$$

The furfural concentration in transformer oil could be obtained using this equation. Therefore, the minimum concentration of dissolved furfural in transformer oil was about 1.06 mg/L according to the established detection method. After the concentration of furfural was determined, the degree of polymerization of insulating paper in transformer according to the following formula could be calculated, so the aging state of a transformer could be determined according to the following (Table 2) [46,47]:

$$\log_{10}(C_{furfural}) = 1.56 - 0.0033DP \quad (2)$$

where $C_{furfural}$ is the concentration of furfural, and $DP$ is the degree of polymerization of the insulating paper in transformer. Therefore, SERS technology is very promising for detecting the concentration of dissolved furfural in transformer oil and further diagnosing the aging state of a transformer.

**Table 2.** The relationship between transformer operation state and the degree of polymerization (DP) [47].

| DP Value | Estimated Percentage of Remaining Life | Suggested Interpretation |
|---|---|---|
| 800 | 100 | Normal Ageing Rate |
| 700 | 90 | |
| 600 | 79 | |
| 500 | 66 | Accelerated Ageing Rate |
| 400 | 50 | |
| 380 | 46 | |
| 360 | 42 | |
| 340 | 38 | Excessive Ageing Danger Zone |
| 320 | 33 | |
| 300 | 29 | |
| 280 | 24 | High Risk of Failure |
| 260 | 19 | |
| 240 | 13 | End of expected life of paper |
| 220 | 7 | |
| 200 | 0 | |

## 4. Conclusions

In order to realize in situ detection of furfural dissolved in transformer oil, silver nanoparticles were self-assembled on the surface of gold film with P-aminophenylthiophenol (PATP) as a coupling agent. R6G was used as the probe molecule to test the enhancement effect. In the process of preparing enhanced substrates, the effects of PATP concentration, deposition time in PATP, and immersion time in silver sol on the reinforcing properties of substrates were studied. By optimizing the molecular concentration, molecular deposition time, and silver sol deposition time of PATP, the nanoparticles were made more uniform and compact, and an enhanced substrate with rich hot spots was obtained. The optimum substrate was developed, and surface-enhanced Raman spectroscopy detection of trace furfural dissolved in transformer oil was realized. The results showed that the substrate prepared under the conditions of 0.1 mol/L PATP, 5 h deposition in PATP, and 12 h immersion in silver sol had the best reinforcement effect, which was uniform and compact particle arrangement and no particle clusters. By use of this substrate, the minimum detectable concentration of furfural in transformer oil was about 1.06 mg/L, which provides a new method for fast and nondestructive detection of transformer aging diagnosis.

**Supplementary Materials:** The following are available online at http://www.mdpi.com/2079-4991/9/1/17/s1, Figure S1: SERS of furfural in transformer oil at 10 acquisition points on a single substrate; Figure S2: SERS of furfural in transformer oil on six substrates.

**Author Contributions:** Writing the manuscript: H.S.; Design of the work and revising the manuscript: W.C., F.W., Y.H.; Acquisition of data: H.S., L.D., C.Z.; Revising the manuscript: H.S., S.Z.; Analysis of data: W.Z., J.Z.

**Funding:** This work was supported by the National Natural Science Foundation special fund joint fund project (U1766217), the science and technology project of the national Power Grid Corp (SGS00200FCJS1700228), and the construction plan of the innovation team at the Colleges of Chongqing (CXTDX201601001).

**Conflicts of Interest:** The authors declare no conflict of interest.

## References

1. Shroff, D.H.; Stannett, A.W. A review of paper aging in power transformers. *IEE Proc. C Gener. Transm. Distrib.* **1985**, *132*, 312–319. [CrossRef]
2. Lundgaard, L.E.; Hansen, W.; Linhjell, D.; Painter, T.J. Aging of oil-impregnated paper in power transformers. *IEEE Trans. Power Deliv.* **2004**, *19*, 230–239. [CrossRef]
3. Arshad, M.; Islam, S.M.; Khaliq, A. Power transformer aging and life extensio. In Proceedings of the 2004 International Conference on Probabilistic Methods Applied to Power Systems, Ames, IA, USA, 12–16 September 2004.
4. Shiri, A.; Gholami, A.; Shoulaie, A. Investigation of the ambient temperature effects on transformer's insulation life. *Electr. Eng.* **2011**, *93*, 193–197. [CrossRef]
5. Cleary, G.P.; Judd, M.D.; Farish, O. Investigation of partial discharge current pulses in transformer insulating oil. In Proceedings of the International Conference on Advances in Processing, Testing and Application of Dielectric Materials (APTADM'2001), Wroclaw, Poland, 17 September 2001.
6. Fofana, I.; Bouaïcha, A.; Farzaneh, M. Characterization of aging transformer oil–pressboard insulation using some modern diagnostic techniques. *Eur. Trans. Electr. Power* **2013**, *21*, 1110–1127. [CrossRef]
7. Lelekakis, N.; Wijaya, J.; Martin, D.; Susa, D. The effect of acid accumulation in power-transformer oil on the aging rate of paper insulation. *IEEE Electr. Insul. Mag.* **2014**, *30*, 19–26. [CrossRef]
8. Oommen, T.V.; Prevost, T.A. Cellulose insulation in oil-filled power transformers: Part II maintaining insulation integrity and life. *IEEE Electr. Insul. Mag.* **2006**, *22*, 5–14. [CrossRef]
9. Stebbins, R.D.; Myers, D.S.; Shkolnik, A.B. Furanic compounds in dielectric liquid samples: Review and update of diagnostic interpretation and estimation of insulation ageing. In Proceedings of the IEEE International Conference on Properties and Applications of Dielectric Materials, Nagoya, Japan, 1–5 June 2003; pp. 921–926.
10. Unsworth, J.; Mitchell, F. Degradation of electrical insulating paper monitored with high performance liquid chromatography. *IEEE Trans. Electr. Insul.* **1990**, *25*, 737–746. [CrossRef]
11. Pablo, A.D. *Furfural and Ageing: How Are They Related Insulating Liquids*; IET: Leatherhead, UK, 1999; pp. 1–4.
12. Yang, L.; Lin, Y.; Liao, R.; Zhao, X.; Sun, W.; Zhang, Y. Effects of temperature and aging on furfural partitioning in the oil-paper system of power transformers. *IEEE Trans. Dielectr. Electr. Insul.* **2016**, *23*, 1393–1401. [CrossRef]
13. Łojewski, T.; Sawoszczuk, T.; Łagan, J.M.; Zięba, K.; Barański, A.; Łojewska, J. Furfural as a marker of cellulose degradation. A quantitative approach. *Appl. Phys. A* **2010**, *100*, 873–884. [CrossRef]
14. Arshad, M.; Islam, S.M. Significance of cellulose power transformer condition assessment. *IEEE Trans. Dielectr. Electr. Insul.* **2011**, *18*, 1591–1598. [CrossRef]

15. Zhang, C.H.; Macalpine, J.M.K. Furfural Concentration in Transformer Oil as an Indicator of Paper Ageing, Part 1: A Review. In Proceedings of the IEEE Pes Power Systems Conference and Exposition, Atlanta, GA, USA, 29 October–1 November 2006; pp. 1088–1091.
16. Dilleen, J.W.; Lawrence, C.M.; Slater, J.M. Determination of 2-furaldehyde in transformer oil using flow injection with pulsed amperometric detection. *Analyst* **1996**, *121*, 755–759. [CrossRef]
17. Wang, R.; Huang, X.; Wang, L. Facile electrochemical method and corresponding automated instrument for the detection of furfural in insulation oil. *Talanta* **2016**, *148*, 412–418. [CrossRef] [PubMed]
18. Pradhan, M.K.; Ramu, T.S. On the estimation of elapsed life of oil-immersed power transformers. *IEEE Trans. Power Deliv.* **2005**, *20*, 1962–1969. [CrossRef]
19. Ministry of Power Industry. *DL/T 596-1996, Electric Power Equipment Preventive Test Regulation*; Ministry of Power Industry: Asinel, Spain, 1996.
20. Jobstl, D.; Husoy, T.; Alexander, J.; Bjellaas, T.; Leitner, E.; Murkovic, M. Analysis of 5-hydroxymethyl-2-furoic acid (HMFA) the main metabolite of alimentary 5-hydroxymethyl-2-furfural (HMF) with HPLC and GC in urine. *Food Chem.* **2010**, *12*, 814–818. [CrossRef]
21. Ahmadi, F.; Zarrin, L.; Sharifnia, S.; Hosseini, S.N. Analyzed and Proposed Mechanism of Photo-catalytic Degradation of Furfural at TiO2 Nano-particles by HPLC-UV and LC-MASS Methods. *J. Liq. Chromatogr. Relat. Technol.* **2014**, *37*, 1750–1762. [CrossRef]
22. Zhang, Y.Y.; Song, Y.; Quan-Hong, L.I. A Review on Formation Mechanism, Determination and Safety Assessment of Furfural and 5-Hydroxymethylfurfural (HMF) in Foods. *Food Sci.* **2012**, *33*, 275–280.
23. Chen, W.; Shi, H.; Wan, F.; Wang, P.; Gu, Z.; Li, W.; Ke, L.; Huang, Y. Substrate influence on the polarization dependence of SERS in crossed metal nanowires. *J. Mater. Chem. C* **2017**, *5*, 7028–7034. [CrossRef]
24. Fu, W.; Shi, H.; Chen, W.; Gu, Z.; Du, L.; Wang, P.; Wang, J.; Huang, Y. Charge Transfer Effect on Raman and Surface Enhanced Raman Spectroscopy of Furfural Molecules. *Nanomaterials* **2017**, *7*, 210.
25. Zang, Y.; Shi, H.; Huang, Y.; Zeng, X.; Pan, L.; Wang, S.; Wen, W. SERS Polarization Dependence of Ag Nanorice Dimer on Metal and Dielectric Film. *Chem. Phys. Lett.* **2017**, *684*, 373–377. [CrossRef]
26. Li, W.; Jia, Y.; Liu, A.; Sun, X.; Su, X.; Zhang, X.; Li, K.; Huang, Y. Effects of substrate and polarization on plasmon driven surface catalysis in nanowire-film hybrid system. *Superlattices Microstruct.* **2016**, *100*, 886–891. [CrossRef]
27. Huang, Y.; Fang, Y.; Zhang, Z.; Zhu, L.; Sun, M. Nanowire-supported plasmonic waveguide for remote excitation of surface-enhanced Raman scattering. *Light-Sci. Appl.* **2014**, *3*, e199. [CrossRef]
28. Somekawa, T.; Fujita, M.; Izawa, Y.; Kasaoka, M.; Nagano, Y. Furfural analysis in transformer oils using laser raman spectroscopy. *IEEE Trans. Dielectr. Electr. Insul.* **2015**, *22*, 229–231. [CrossRef]
29. Chen, W.; Gu, Z.; Zou, J.; Wan, F.; Xiang, Y. Analysis of furfural dissolved in transformer oil based on confocal laser Raman spectroscopy. *IEEE Trans. Dielectr. Electr. Insul.* **2016**, *23*, 915–921. [CrossRef]
30. Fleischmann, M.; Graves, P.R.; Robinson, J. Enhanced and normal Raman scattering from pyridine adsorbed on rough and smooth silver electrodes. *J. Electroanal. Chem. Interfacial Electrochem.* **1985**, *182*, 73–85. [CrossRef]
31. Li, Z.; Hao, F.; Huang, Y.; Fang, Y.; Nordlander, P.; Xu, H. Directional Light Emission from Propagating Surface Plasmons of Silver Nanowires. *Nano Lett.* **2009**, *9*, 4383–4386. [CrossRef]
32. Huang, Y.; Fang, Y.; Sun, M. Remote Excitation of Surface-Enhanced Raman Scattering on Single Au Nanowire with Quasi-Spherical Termini. *J. Phys. Chem. C* **2011**, *115*, 3558–3561. [CrossRef]
33. Sharma, B.; Frontiera, R.R.; Henry, A.I.; Ringe, E.; Van Duyne, R.P. SERS: Materials, applications, and the future. *Mater. Today* **2012**, *15*, 16–25. [CrossRef]
34. Kneipp, K.; Wang, Y.; Kneipp, H.; Perelman, L.T.; Itzkan, I.; Dasari, R.R.; Feld, M.S. Single Molecule Detection Using Surface-Enhanced Raman Scattering (SERS). *Phys. Rev. Lett.* **1997**, *78*, 1667. [CrossRef]
35. Stetciura, I.Y.; Yashchenok, A.; Masic, A.; Lyubin, E.V.; Inozemtseva, O.A.; Drozdova, M.G.; Markvichova, E.A.; Khlebtsov, B.N.; Fedyanin, A.A.; Sukhorukov, G.B.; et al. Composite SERS-based satellites navigated by optical tweezers for single cell analysis. *Analyst* **2015**, *140*, 4981–4986. [CrossRef]
36. Ilkhani, H.; Hughes, T.; Li, J.; Zhong, C.J.; Hepel, M. Nanostructured SERS-electrochemical biosensors for testing of anticancer drug interactions with DNA. *Biosens. Bioelectron.* **2016**, *80*, 257–264. [CrossRef]
37. Wang, H.; Liu, T.; Huang, Y.; Fang, Y.; Liu, R.; Wang, S.; Wen, W.; Sun, M. Plasmon-driven surface catalysis in hybridized plasmonic gap modes. *Sci. Rep.* **2014**, *4*, 7087. [CrossRef]

38. Vo-Dinh, T. SERS chemical sensors and biosensors: New tools for environmental and biological analysis. *Sens. Actuators B Chem.* **1995**, *29*, 183–189. [CrossRef]
39. Wang, P. Graphene-Plasmonic Hybrid Platform for Label-Free SERS Biomedical Detection. Ph.D. Thesis, UCLA, Los Angeles, CA, USA, 2015.
40. Lee, C.H.; Tian, L.; Singamaneni, S. Paper-based SERS swab for rapid trace detection on real-world surfaces. *Acs Appl. Mater. Interfaces* **2010**, *2*, 3429. [CrossRef]
41. Anema, J.R.; Li, J.F.; Yang, Z.L.; Yang, Z.L.; Li, S.B.; Zhou, X.; Fan, F.R.; Zhang, W.; Zhou, Z.Y.; Wu, D.Y.; et al. Shell-Isolated Nanoparticle-Enhanced Raman Spectroscopy: Expanding the Versatility of Surface-Enhanced Raman Scattering. *Annu. Rev. Anal. Chem.* **2011**, *4*, 129. [CrossRef]
42. Rodal-Cedeira, S.; Montes-García, V.; Polavarapu, L.; Solís, D.M.; Heidari, H.; La Porta, A.; Angiola, M.; Martucci, A.; Taboada, J.M.; Obelleiro, F.; et al. Plasmonic Au@Pd nanorods with boosted refractive index susceptibility and SERS efficiency: A multifunctional platform for hydrogen sensing and monitoring of catalytic reactions. *Chem. Mater.* **2016**, *28*, 9169–9180. [CrossRef]
43. GómezGraña, S.; FernándezLópez, C.; Polavarapu, L.; Salmon, J.B.; Leng, J.; Pastoriza-Santos, I.; Pérez-Juste, J. Gold Nanooctahedra with Tunable Size and Microfluidic-Induced 3D Assembly for Highly Uniform SERS-Active Supercrystals. *Chem. Mater.* **2015**, *27*, 8310–8317. [CrossRef]
44. Jeong, C.; Kim, H.-M.; Park, S.Y.; Cha, M.G.; Park, S.J.; Kyeong, S.; Pham, X.H.; Hahm, E.; Ha, Y.; Jeong, D.H.; et al. Highly Sensitive Magnetic-SERS Dual-Function Silica Nanoprobes for Effective On-Site Organic Chemical Detection. *Nanomaterials* **2017**, *7*, 146. [CrossRef]
45. Lee, P.C.; Meisel, D. Adsorption and surface-enhanced Raman of dyes on silver and gold sols. *J. Phys. Chem.* **1982**, *86*, 3391–3395. [CrossRef]
46. Lin, Y.; Yang, L.; Liao, R.; Sun, W.; Zhang, Y. Effect of oil replacement on furfural analysis and aging assessment of power transformers. *IEEE Trans. Dielectr. Electr. Insul.* **2015**, *22*, 2611–2619. [CrossRef]
47. Mehta, D.; Jariwala, H. Predication of Life of Transformer insulation by developing Relationship between Degree of Polymerization and 2- Furfural. *Int. J. Sci. Eng. Res.* **2012**, *3*, 1–4.

nanomaterials

MDPI

Article

# Detection of Circulating Tumor Cells Using Membrane-Based SERS Platform: A New Diagnostic Approach for 'Liquid Biopsy'

**Agnieszka Kamińska [1,*], Tomasz Szymborski [1], Evelin Witkowska [1], Ewa Kijeńska-Gawrońska [2], Wojciech Świeszkowski [2], Krzysztof Niciński [1], Joanna Trzcińska-Danielewicz [3] and Agnieszka Girstun [3]**

1 Institute of Physical Chemistry, Polish Academy of Sciences, Kasprzaka 44/52, 01-224 Warsaw, Poland; tszymborski@gmail.com (T.S.); ewitkowska@ichf.edu.pl (E.W.); knicinski@ichf.edu.pl (K.N.)

2 Faculty of Materials Science and Engineering, Warsaw University of Technology, Wołoska 141, 02-507 Warsaw, Poland; ewakijenska@gmail.com (E.K.-G.); wojciech.swieszkowski@pw.edu.pl (W.Ś.)

3 Department of Molecular Biology, Institute of Biochemistry, Faculty of Biology, University of Warsaw, Miecznikowa 1, 02-096 Warsaw, Poland; jtd@biol.uw.edu.pl (J.T.-D.); agirstun@biol.uw.edu.pl (A.G.)

* Correspondence: akamin@ichf.edu.pl

Received: 7 January 2019; Accepted: 26 February 2019; Published: 5 March 2019

**Abstract:** The detection and monitoring of circulating tumor cells (CTCs) in blood is an important strategy for early cancer evidence, analysis, monitoring of therapeutic response, and optimization of cancer therapy treatments. In this work, tailor-made membranes (MBSP) for surface-enhanced Raman spectroscopy (SERS)-based analysis, which permitted the separation and enrichment of CTCs from blood samples, were developed. A thin layer of SERS-active metals deposited on polymer mat enhanced the Raman signals of CTCs and provided further insight into CTCs molecular and biochemical composition. The SERS spectra of all studied cells—prostate cancer (PC3), cervical carcinoma (HeLa), and leucocytes as an example of healthy (normal) cell—revealed significant differences in both the band positions and/or their relative intensities. The multivariate statistical technique based on principal component analysis (PCA) was applied to identify the most significant differences (marker bands) in SERS data among the analyzed cells and to perform quantitative analysis of SERS data. Based on a developed PCA algorithm, the studied cell types were classified with an accuracy of 95% in 2D PCA to 98% in 3D PCA. These results clearly indicate the diagnostic efficiency for the discrimination between cancer and normal cells. In our approach, we exploited the one-step technology that exceeds most of the multi-stage CTCs analysis methods used and enables simultaneous filtration, enrichment, and identification of the tumor cells from blood specimens.

**Keywords:** surface-enhanced Raman spectroscopy; circulating tumor cells (CTC); prostate cancer (PC3); cervical carcinoma (HeLa); label-free detection

## 1. Introduction

Circulating tumor cells (CTCs) are shed from the cancerous mass of the primary tumor cells and may get into the peripheral blood and then transfer to distant tissues causing expansion of the cancer in the metastasis form [1]. CTCs are a crucial source of genetic material for clinical analysis, e.g., tumor diagnostics, and selection and monitoring of cancer therapy [2–6]. Tissue biopsies are invasive and expensive, therefore the detection and characterization of tumors based on CTC analysis in a sample of peripheral blood, known as 'liquid biopsy', is the subject of special interest. A variety of techniques have been developed for isolation, capturing, and analysis of circulating tumor cells. The most widely used approaches include polymerase chain reaction (PCR) and reverse transcription PCR

(RT-PCR) [7], immunofluorescence [8–10], fluorescence-based cytometry [11], fluorescence scanning microscopy [12–14], and label-free biochemical separation [15]. Recently, the microfluidic devices based on miniaturized nanomaterials and microfluidic reactions improved the sensitivity of detection and enabled continuous single-cell analysis [16,17]. The current 'gold standard method' of CTC detection and enumeration is the CellSearch system (Veridex LLC) for the analysis of CTCs in metastatic breast [18], prostate [19], and colon cancer patients [20]. In this method ferromagnetic beads are coated with antibodies against the epithelial cellular adhesion molecule (EpCAM), then immunostained with fluorescently labeled anti-cytokeratin (CK, an epithelial intermediate filament), anti-CD45 (a membrane antigen expressed by leucocytes) antibodies, then stained with DAPI (4′,6-diamidino-2-phenylindole, a nuclear stain), and finally counted by automated cell image capture and analysis even from a 7.5 mL blood sample [21].

However, it should also be highlighted that all the above described techniques require the use of specific molecular markers for CTCs detection and are time-consuming and/or expensive, which limits these methods for routine clinical analysis. For example, RT-PCR enables only the examination of a limited number of genes at the same time and does not permit the morphological analysis of cells in subsequent tests [22]. Moreover, the ability of this technique to detect multiple cancer markers might by hindered by the lack of appropriate tumor marker expression. In all the above mentioned methods, the outcomes have to be confirmed by histopathological examination of the tissue collected during biopsy. In order to reduce high medical costs of biopsy, we attempted to develop a new approach to cancer diagnosis based on the Raman technique.

Surface-enhanced Raman spectroscopy (SERS) is a highly sensitive and specific method that allows for the detection and characterization of various compounds through their capability to generate specific molecular fingerprint signals. SERS spectroscopy reveals the huge enhancement of Raman scattering signals of molecules adsorbed on specially prepared metallic nanostructures, usually made of silver, gold, or copper [23]. The two main mechanisms, electromagnetic (EM) and chemical (CT), can increase Raman peaks intensities by 9–14 orders of magnitude relative to normal Raman spectroscopy, which gives the possibility of single molecule detection [24]. The huge enhancement factor (huge sensitivity), high selectivity, possibility of label-free, rapid, and non-destructive analysis leads to an increase in the practical applications of this technique, especially in biomedical and analytical studies.

Recently, various studies have shown the capability of SERS in tumor cells identification [25–27]. Zhang et al. presented a novel strategy based on the nitrocellulose membrane and SERS imaging method, which can be used for both CTC enrichment and detection [28]. Jun et al. [29] developed silica-encapsulated magnetic nanoparticles (MNPs) with unique properties for cancer cell targeting and identification. Wang et al. used SERS-active nanoparticles modified with epidermal growth factor (EGF) peptide as a targeting ligand for efficient CTC detection in blood plasma [30]. Wen et al. successfully developed a method which has an ability to quickly respond and can be used for CTC capturing and detection. The CTC capturing and detection efficiency was also proved via real blood samples from clinical subjects using magnetic nanospheres [31]. Shi et al. [32] reported detection of cervical carcinoma (HeLa) cells using the designed folate-conjugated SERS-active nanoparticles and the magnetic tapping strategy. Krafft et al. [33] presented a microfluidic chip combined with optical tweezers to collect the normal Raman spectra of circulating cells.

Most of the presented SERS based methods of CTC analysis require an enrichment step of a few CTCs from the blood and an extra labelling strategy for preparation, e.g., Raman reporter or peptides encoded NPs and/or tumor cells pre-labelled with NPs [29], which increases the cost and time of analysis.

Currently, various methods have been explored to optimize the properties of SERS platforms in terms of size, shape, and composition of used plasmonic nanostructures [34–37]. Even though many efficient SERS supports have been developed [38,39], there are some disadvantages associated with them, e.g., low stability over time, and inability to perform in situ measurements and sample mapping. To avoid these problems, new production strategies of SERS platforms should be developed in order to introduce SERS techniques to standard biomedical and analytical applications. We have demonstrated that polymer mats covered via the PVD (physical vapor deposition) technique with gold or gold–silver alloy may work as a very efficient SERS platform for the identification of bacteria from environmental, food, and human body fluid samples [40,41].

In our approach we offer a simple method that is optimal for CTC isolation, enrichment, detection, and molecular analysis. We have elaborated the procedure of membrane-based SERS platforms (MBSP) with appropriate pore sizes that permit the separation and enrichment of the prostate cancer cell line (PC3), cervical carcinoma cell line (HeLa), and leucocytes—an example of healthy cells in the blood samples. By covering tailor-made SERS substrates with a thin layer of Ag–Au alloy, the high enhancement of Raman signals of all studied CTCs was achieved.

Additionally, to improve the efficiency of discrimination between cancer and blood cells, principal component analysis (PCA) was adopted. PCA is one of the most commonly used methods in the classification of SERS data, which enables the identification of the spectral differences among the studied samples, extracting the characteristic marker bands and biochemical information from SERS spectra. The presented strategy may lead to development of new tools in CTC therapy.

## 2. Experimental Section

The prostate cancer (PC3) and cervical carcinoma (HeLa) cell lines used in this work were obtained from the European Collection of Cell Cultures (ECACC), Sigma-Aldrich (St Louis, MO, USA). PC3 and HeLa cells were cultured in RPMI-1640 and DMEM media, respectively. Both media were supplemented with 10% fetal bovine serum (FBS), streptomycin (100 µg/mL), and penicillin (100 U/mL). The cell cultures were cultivated at 37 °C, in a humidified atmosphere of 5% $CO_2$. During experiments the cancer cells were cultured in 25 $cm^2$ cell culture flasks. After reaching 80% of confluence, the cells were suspended in phosphate buffer saline (PBS) buffer and trypsinized (0.05% trypsin, 0.02% EDTA solution). Subsequently, the cells were collected, centrifuged at 250 × g for 5 min at room temperature, re-suspended in PBS and centrifuged repeatedly. After the last centrifugation the sample containing 20 µL of PBS was obtained and stored on ice. All the chemical reagents were obtained from Sigma-Aldrich (St. Louis, MO, USA).

During the preparation of prostate cancer (PC3) and cervical carcinoma (HeLa) cell lines for SERS experiments we used concentrations reflecting the population of cancer cells in metastasis. The initial concentration (after the cultivation step) of cancer cells in PBS was $0.44 \times 10^6$ cells/mL for PC3 and *ca.* $10^6$ cells/mL for HeLa, respectively, and was further diluted to the final concentration of *ca.* 40 cells in 1mL of blood.

The human blood samples derived from ten healthy volunteers, available courtesy of the Regional Blood Center (Warsaw, Poland), were used in our studies. An informed consent was obtained from all subjects (healthy volunteers). The performance of all experiments was in agreement with the institutional guidelines and relevant laws and approved by the Ethics and Bioethics Committee of Cardinal Stefan Wyszyński University (Warsaw, Poland).

### 2.1. Leucocyte Isolation

Whole blood samples were lysed with five volumes of hypotonic erythrocyte lysis buffer (RBCL, A&A Biotechnology, Gdynia, Poland) from at least 20 mL of peripheral blood. After 15 min of incubation on ice and centrifugation (3000× *g*), the plasma-free leukocytes were re-suspended in PBS solution at a concentration of $2 \times 10^7$ cells/mL.

### 2.2. Fabrication of Membrane-Based SERS Platforms (MBSP) via Electrospinning

Poly(l-lactic acid)-co-poly(ε-caprolactone) (P(LLA-CL)) with a ratio of 70:30 used for fiber fabrication was purchased from Evonik (Witten, Germany). Prior to electrospinning, two types of P(LLA-CL) solutions with concentrations of 10% and 14% (w/v) were prepared by dissolving polymer powder (crystals) in 1,1,1,3,3,3-hexafluoro-2-propanol (Fluorochem, Hadfield, Derbyshire,

UK) and stirring overnight in ambient conditions. The solutions were then individually placed in 10 mL plastic syringes.

The electrospinning process was carried out under optimized conditions with the use of NANON-01A (MECC Co., Ltd.; Fukuoka, Japan). Two 27 G steel needles were connected with syringes using polytetrafluoroethylene (PTFE) tubes, fixed to the moving head with a constant linear velocity of 100 mm/min, and attached to high voltage of 15 kV provided by a built-in power supply. The distance between the needles was 25 mm, and the tips-to-collector distance was set at 150 mm. The feed rate for both solutions was 1.0 mL/h. The working width of the moving head was 100 mm. The nanofibrous meshes were collected on the aluminum covered steel plate and dried in a vacuum drier for 24 h (25 °C, 50 mb).

### *2.3. Sputtering of the Thin Layer of SERS Active Metal*

The PVD device (Leica, EM MED020, Heerbrugg, Switzerland) was applied to sputter 40 nm of Ag:Au alloy directly on the polymer fibers. No adhesion layer, i.e., chromium or titanium, was used between the polymer and the Ag:Au alloy layer. The sputtering conditions were: current of 25 mA and pressure of $10^{-2}$ mbar.

## **3. Instrumentation**

### *3.1. SERS Measurements*

Measurements were carried out with a Renishaw inVia Raman system (Wotton-under-Edge, Gloucestershire, UK) equipped diode laser emitting a 785 nm laser line. The light from the lasers was passed through a line filter and focused on a sample mounted on an X–Y–Z translation stage with a 50× microscope objective, NA = 0.25. The beam diameter was approximately 2.5 µm. The laser power at the sample was 1.5 mW. The experiments were performed at ambient conditions using a back-scattering geometry. The spectroscopic maps were acquired by collecting SERS spectra over the previously defined range (36 × 21 $\mu m^2$) at each point on a grid with 3 µm spacing using an automated microscope stage. Typically, 25 SERS spectra for each cell type were acquired. Each spectrum was measured for 30 s.

### *3.2. SEM Measurements*

SEM images were acquired from the FEI Nova NanoSEM 450 instrument (Hillsboro, OR, USA) operating at an accelerating voltage of 10 kV and under high vacuum.

### *3.3. Chemometrics—Principal Component Analysis*

Principal component analysis (PCA) is a multivariate procedure that can reduce the dimensionality of original raw data to several principal components (PCs). It is an effective technique that gives the possibility to categorize SERS spectra that are readily distinguishable via visual empirical analysis. The calculated PCs contain the most significant information from the whole introduced data set. The PCA was performed using the commercial Unscrambler® software (CAMO software AS, version 10.3, Oslo, Norway). The SERS data of all analyzed cells (leucocytes, HeLa, and PC3 cells) were optimized for PCA using the following steps: (i) smoothing with a Savitzky–Golay filter (Oslo, Norway), (ii) background correction (concave rubber band correction; the number of baseline points was 34 and the number of iterations was 10), and (iii) normalization using OPUS software (Bruker Optic GmbH, 2012 version, Ettlingen, Germany). The PCA was completed based on the NIPLAS algorithm, validation (random with 20 segments), significance 0.05, and a SERS spectra number of 120.

## 4. Results and Discussion

### 4.1. Preparation of the SERS Platform

In this study, we present a novel SERS platform prepared with the use of the electrospinning technique [42] for the label-free analysis of CTCs in blood samples. Fabrication of MBSP consisted of two steps:

(i) Electrospinning of polymer mats with desired parameters, i.e., diameter of the polymer fibers and diameter of the pores;

(ii) Sputtering of thin (usually tens of nanometers) layer of SERS-active metal, e.g., gold, silver, or their alloy via the PVD method.

The basic scheme of the utilized method for step (i) is shown in Figure 1.

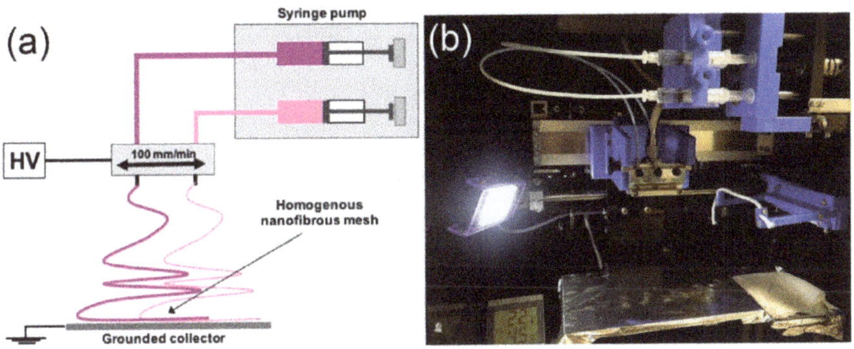


**Figure 1.** (**a**) Basic layout of the setup utilized for electrospinning, which consists of a high voltage power supply (HV), two syringe pumps, and a grounded collector; (**b**) a photo of experimental setup used in experiments.

To prepare SERS-active platforms there was a need to coat polymer mats with metal NPs or metal islands. In the case of thin metal islands, it could be done by PVD or vacuum evaporation. In our study the layer of Ag:Au alloy was sputtered on the polymer fibers via the PVD method. The Ag:Au alloy ensured the combination of very high enhancement of the Raman signal provided by Ag with the chemical stability offered by Au [43]. In order to create a platform that provided optimal enhancement of the Raman signal, three different thicknesses of Ag:Au alloys (20, 40, and 80 nm) were tested.

The Ag:Au alloy layer of 20 nm deposited on the polymer mat was too thin to cover the platform and thus to obtain the SERS signal of *p*-MBA (*p*-mercaptobenzoic acid) or tumor cells. As a result, the recorded SERS spectra were derived from the polymer. The polymer mat covered with 40 nm of Ag:Au alloy showed the greatest SERS enhancement. No SERS signals from the polymers were observed. A similar level of enhancement was achieved for the 80 nm Ag:Au layer. Therefore, in the present study, the polymer mats covered with the 40 nm layer were used for all experiments as the most cost-effective. Moreover, the process of sputtering of 40 nm of Ag:Au alloy took only 4 min compared to 8 min for the 80 nm layer.

The morphology of the created SERS substrates named Au:Ag/MBSP SERS was examined by scanning electron microscopy (SEM). The SEM images of (P(LLA-CL)) covered with 40 nm of the Ag:Au alloy layer are presented in Figure 2 at smaller and larger magnifications, respectively. As can be seen in Figure 2a, the arrangement of fibers with a diameter of *ca.* 1.5 μm within mats was irregular with the slots between fibers of *ca.* 15 μm that were small enough not to let the tumor cells (with the diameter of 20–28 μm) pass through the (P(LLA-CL)) mat. Additionally, the SEM image shown in Figure 2b reveals that the obtained layer of Ag:Au consisted of semi-spheres with diameters ranging from 40 to 55 nm, and their size was responsible for the enhancement factor of the presented Au:Ag/PBSP SERS substrates, and determined the SERS efficiency of these surfaces.

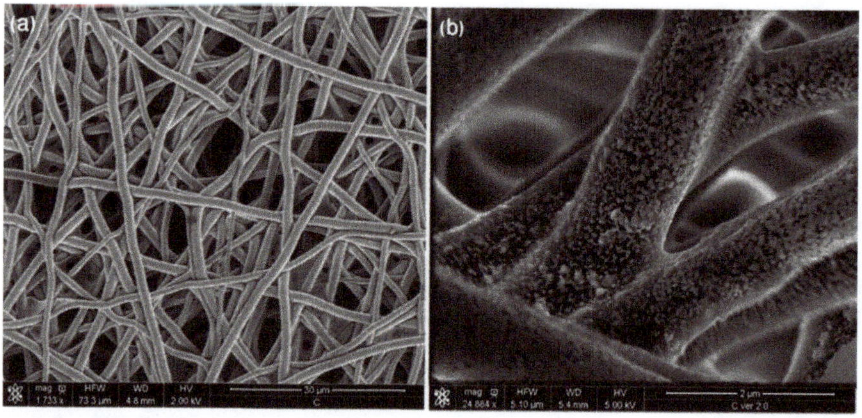


**Figure 2.** SEM images of electrospun polymer mat coated with gold layer (40 nm) at (**a**) lower and (**b**) higher magnification.

The SERS platform designed in such a way worked also as a filter, which allowed the separation of circulating prostate cancer (PC3) and cervical carcinoma (HeLa) cells re-suspended in human blood plasma at a concentration of about 40 cells in 1mL of blood.

The main advantage of the proposed method is the fact that it does not require use of separate techniques to perform filtration, enrichment, and examination of tumor cells circulating in blood. Additionally, by combining these three basic steps in detecting cancer cells in one single process, the transfer of the cells from one place/method to another is eliminated. Therefore, the proposed strategy prevents contamination of the samples and disintegration of cell structures, and leads to improvement of the accuracy of analysis and reduction of the time of analysis.

### 4.2. SERS Investigations of Circulating Tumor Cells

The Au:Ag/MBSP SERS platform worked both as a filter and as an efficient SERS support and allowed for: (i) separation of studied cells from the complex blood sample due to the sizes of MBSP pores (Figure 2) and sizes of particular blood components (Table S1), and (ii) enrichment of circulating tumor cells within a small and defined area of the SERS substrate. Figure 3a illustrates the experimental setup used for the detection of studied cells whereas Figure 3b demonstrates the filtration process.

As the amount of single CTCs in peripheral blood is small [44], the highly-efficient cell enrichment and single cell capturing were essential for further cell examination.

The proposed concept based on spiking blood samples obtained from healthy donors with a known number of HeLa and PC3 cells (40 cells in 1 mL of blood) may have in the future a practical potential in medicine.

In order to push the sample through the device, a constant pressure of approximate $80 \times 10^3$ Pa was applied. The whole filtration process took about 4 minutes. Since the pores in the Au:Ag/MBSP SERS platform had a diameter of *ca.* 15 µm and the sizes of blood components and analyzed CTCs did not exceed 15 and 28 µm, respectively (see Table S1, Supplementary Materials), the separation of CTCs from other blood components could be performed. The smaller components of blood plasma passed through the Au:Ag/MBSP SERS platform whilst the largest CTCs remained on the surface of the modified (P(LLA-CL)) mat. As mentioned before, the Ag:Au nanostructures present on the polymer mat fibers are responsible for amplification of the Raman signal of CTCs. Therefore, in the next step the spectroscopic fingerprints of captured CTCs were recorded to perform detailed molecular analysis and identification of studied cells.

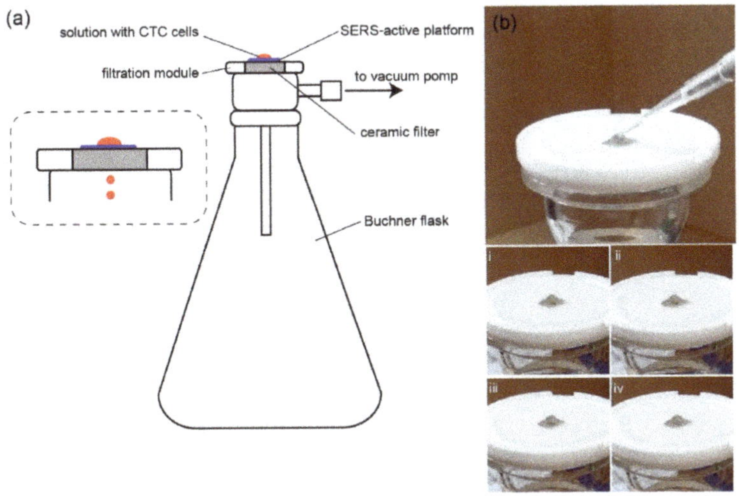


**Figure 3.** (**a**) The scheme of capturing circulating tumor cells (CTCs) from the blood sample. The system involves: a vacuum pump, Buchner flask, and filter funnel. The surface-enhanced Raman spectroscopy (SERS) platform was placed on the filter funnel and a droplet of blood spiked with CTCs was put on the platform. After turning on the pump, the liquid was sucked through the mat to the flask, whereas the CTCs remained on the surface of the SERS platform. (**b**) Filtration process of the fluid with the CTCs. The setup consists of a ceramic filter and SERS-active platform placed in the very center. After pipetting a small amount of fluid (top) with CTCs, the vacuum pump is turned on and the blood passes through the mat and the ceramic filter to the Büchner flask, whereas the CTCs stay on the SERS-active platform (steps i–iv).

In order to collect the reference spectra of all studied cells (PC3, HeLa, and leucocytes as an example of healthy cells) the SERS measurements were performed directly from pre-cultures (see Figure S1, Supplementary Materials). Table 1 presents the main SERS bands observed in analyzed cell spectra and the corresponding bands assignments.

Figure 4 depicts the SERS spectra of the cells isolated from blood samples using the Au:Ag/MBSP SERS platform according to the procedure discussed above. As can be seen these spectra showed differences in the position of some bands and their relative intensities. However, the common bands corresponded to the main components of the eukaryotic cell [45]: nucleic acids, proteins, and lipids were clearly observed in all SERS spectra.

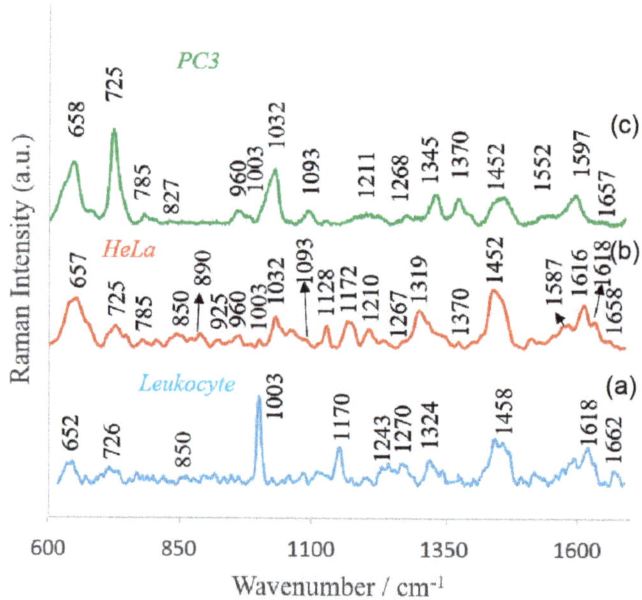


**Figure 4.** Averaged and normalized SERS spectra of (**a**) leucocytes, (**b**) cervical carcinoma (HeLa), and (**c**) prostate cancer (PC3) cells recorded on polymer-based SERS platform. Experimental conditions: excitation at 785 nm, laser power at 1.5 mW, and 45 seconds integration time. Each SERS spectrum was obtained by averaging at least 25 single spectra from different places on the SERS substrate.

All the spectral fingerprints depicted in Figure 4 corresponded with the reference SERS data in Figure S1. In Figure 4 one can observe that the vibrational modes of nucleic acids were present at 785 and 1093 cm$^{-1}$. The week bands around 1268 and 1660 cm$^{-1}$ were characteristic of amide I and amide III bands, respectively. In all recorded SERS spectra there appeared vibrational modes characteristic of phenylalanine (1003 cm$^{-1}$), tyrosine (850 cm$^{-1}$), and tryptophan (725 cm$^{-1}$). As can be observed, the SERS spectrum of particular cells also had their own specific spectral features. For example, the band at 1345 cm$^{-1}$, which corresponded to adenine and guanine, could be seen in PC3 cells, but not in the HeLa cells and leucocytes. Additionally, the relative intensities of some bands could also be used for differentiation of analyzed cells. To make identification of PC3 and HeLa cells, the ratio of the relative intensities of the bands at 658 cm$^{-1}$/725 cm$^{-1}$ could be used. In the SERS spectra of leucocytes, the most prominent bands appeared at 652 cm$^{-1}$ (C–C twist of tyrosine) [46], 726 cm$^{-1}$ (C–S in protein, $CH_2$ rocking, adenine) [47], 1003 cm$^{-1}$ (C–C of phenylalanine) [48], 1170 cm$^{-1}$ (C–H in plane of tyrosine or nucleic acid) [49], 1458 cm$^{-1}$ (nucleic acid nucleotides) [50], and 1618 cm$^{-1}$ ($\nu$(C=C), tryptophan, tyrosine) [51]. All these dissimilarities enabled recognition of circulating cells. The spectroscopic data revealed that the healthy leucocyte cells could be distinguished from tumor cells using bands at 1032 cm$^{-1}$ ($CH_2CH_3$ bending modes of lipids) [51] and 1452 cm$^{-1}$ (structural protein modes of tumors) [52]. The intensive band at 1452 cm$^{-1}$ was assigned to overlapping asymmetric $CH_2$ bending and $CH_2$ scissors vibrations. The bands of phospholipids, elastin, and collagen were also identified in this region [41]. These differences reflected the changes in biochemical pattern of cancer cells (compared to healthy cells) as the result of carcinogenesis. Table 1 depicts all observed SERS bands with their assignments.

**Table 1.** Assignment of SERS bands depicted in Figure 4 [49,53–57].

| Observed SERS Band ($cm^{-1}$) | Protein | Lipids | Nucleic Acid |
|---|---|---|---|
| 652–658 | Tyr (C–C twist) | | |
| 725–730 | Trp | C–N head group choline $(H_3C)_3N+$ | A |
| 785 | | | $PO_2$ symm |
| 827 | | | RNA backbone |
| 850 | Tyr, Pro | | |
| 890 | Structural protein modes of tumors | | |
| 925 | C–C str alpha-helix, Pro, Val | | |
| 960 | $CH_3$ def | $CH_3$ def | |
| 1003 | Phe | | |
| 1030–1032 | | $CH_2CH_3$ bending modes of lipids | |
| 1093 | C–N stretch | CC str chain, C–O str | $PO_2$ symm |
| 1128 | C–N str bk | porphyrin | |
| 1170–1172 | Tyr C–H inplane | | T |
| 1210 | C–$C_6H_5$ str in phenylalanine tyrosine | | |
| 1243 | Amide III (beta sheet) | | |
| 1267–1270 | Amide III ( random coil) | =CH def | |
| 1319 | $CH_3$ def, collagen | $CH_3CH_2$ twist | G |
| 1324 | | | purine bases of DNA |
| 1345 | | | A, G |
| 1370 | | sphingoglycolipids | |
| 1452 | structural protein modes of tumors | | |
| 1458 | | | A, G |
| 1552 | | | A, G |
| 1597–1600 | Phe, Tyr | | |
| 1616–1618 | C≡C str of Tyr and Trp | | |
| 1657–1665 | Amide I | C=C str | |

The reproducibility of recorded SERS signals is a crucial parameter, especially in the terms of real clinical applications. We calculated the reproducibility of the SERS signals of leucocyte, PC3, and HeLa cells (usually 40 SERS spectra for each type of cells measured on the same SERS substrate were considered). The calculated standard deviations (RSDs) were performed for the most prominent bands at 725, 1003, and 1035 $cm^{-1}$. The achieved results were 6.2%, 8%, and 7.5%, respectively (Table S2, Supplementary Materials). Additionally, Figure S2 shows examples of representative SERS spectra of all studied cells collected from different points within the same Au:Ag/MBSP SERS platform. However, the changes in relative intensities of the same bands were observed and were related to the effect of molecular orientation in relation to the polarization of plasmon excitations in the metal substrate.

The SEM images (Figure 5) show different types of cells filtered from blood and immobilized on the SERS-active platforms.

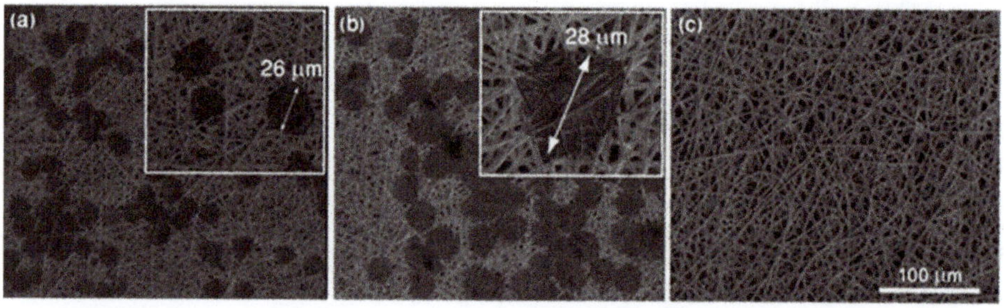


**Figure 5.** SEM images of SERS platforms after filtration of (**a**) HeLa cells, (**b**) prostate cancer PC3 cells, and (**c**) leucocyte cells. White blood cells were not captured on the polymer based SERS-active platform due to their size, which was smaller than the diameter of pores.

It is evident that HeLa and PC3 cells of the size of *ca.* 26 and 28 µm, respectively, could be easily detected on the SERS platform. The leucocytes, which had smaller sizes, passed through the pores and could not be identified by SEM imaging.

### 4.3. Principal Component Analysis

The principal component analysis (PCA) was also performed for the statistical analysis of all studied cells. The data sets including 300 spectra obtained from HeLa, PC3, and leucocyte were analyzed by PCA using the commercial Unscrambler®software (CAMO software AS, version 10.3, Oslo, Norway). It is obvious from looking at Figure 6 that the SERS spectra of different cell types could be easily distinguished (all SERS data were divided into three clusters corresponding to leucocytes (blue), PC3 (green), and HeLa (red) based on the significant PCs (PC-1, PC-2)). We could clearly see that they were distributed into separated regions, which indicated the possibility of differentiation of analyzed cells. The PC-1 and PC-2 were counted in the wavenumber region between 600 and 1700 $cm^{-1}$ and differentiated the cancer cells from normal cells with a sensitivity of 82%.

It should be noticed that the cluster of leucocytes showed a relatively higher homogeneity in comparison to HeLa and PC3 cells, which probably reflects the molecular changes in the structure and biochemical composition of cancer cells [58].

As it can be seen from Figures 4 and 6b many spectral features present in SERS spectra were captured by the main PCs.

The prominent SERS bands at *ca.* 725, 1032, and 1452 $cm^{-1}$ also had the largest weights in the variations and indicated the most significant differences among the three types of analyzed cells. The important contribution to PC-1 also gave bands at *ca.* 1589, 1553, 1470, 1368, and 1380 $cm^{-1}$, analyzed in the previous section. To enhance the sensitivity of differentiation among analyzed cells, a further PCA calculation was made on the limited SERS data (the region between 700 and 750 $cm^{-1}$) where one of the most prominent marker bands at 725 $cm^{-1}$ was observed (Figure S3). In this case, the calculated PC-1 and PC-2 values increased and explained up to 98% of the total variance. This result illustrated that all studied cells were clearly separated into three clusters corresponding to the prostate cancer (PC3), cervical carcinoma (HeLa), and leucocyte cell lines, respectively (Figure S3A).

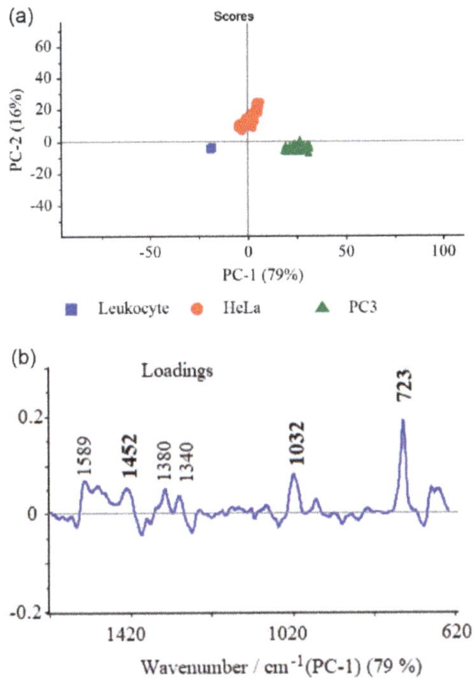


**Figure 6.** (**a**) The score plots of PC-1 versus PC-2 components for differentiation of leucocyte, HeLa, and PC3 cells. PCA was calculated for the whole region (600–1700 $cm^{-1}$); (**b**) PC-1 loading plot.

## 5. Conclusions

Surface-enhanced Raman spectroscopy can be used as an ultrasensitive (at the level of single cell), non-invasive, rapid, and label-free method that provides valuable structural and biochemical analysis of circulating tumor cells. Moreover, Raman spectroscopy coupled with multivariate techniques gives the statistical diagnostic approach for efficient screening of cancer cells.

Our work clearly demonstrates the potential of such a PCA-based SERS method for the direct detection and identification of prostate cancer (PC3) and cervical carcinoma (HeLa) cell lines in blood samples with excellent specificity and sensitivity. We offer a single technology that is optimal for: (i) isolation of CTCs from blood samples, and their (ii) enrichment, (iii) detection, and (iv) molecular analysis.

The developed strategy is based on the tailor-made membrane-based SERS platforms, prepared according to a novel procedure, which permits at that same time the separation, immobilization/enrichment, and enhancement of the week Raman signals of CTCs.

The diagnostic sensitivity of 98% can be achieved for differentiation of PC3, HeLa, and normal cells.

The presented SERS-based strategy for circulating cell detection offers, besides such unique advantages as an ability for rapid and label-free recognition of CTCs with excellent sensitivity and selectivity, also simple sample preparation and cost-effective measurements. Our results indicate that a SERS-based cancer sensor has a great potential to be introduced in a variety of studies conducted on different types of cancer cells, especially from clinical samples.

**Supplementary Materials:** The following are available online at http://www.mdpi.com/2079-4991/9/3/366/s1. Figure S1: Reference SERS spectra of (a) HeLa, (b) PC3, and (c) leucocytes placed directly on SERS substrate, Figure S2: SERS spectra of (a) leucocytes, (b) HeLa, and (c) PC3 cells recorded from different spots (ca. 100) within the same sample. The excitation wavelength was at 785 nm, laser power was 5 mW, and the acquisition time was 60 seconds, Figure S3: (a) First three PC (PC-1, PC-2, and PC-3) scores plot of 3D-PCA calculated for narrow range (700–750 $cm^{-1}$) and (b) loadings plot of the first principal component showing the most prominent marker band at 723 $cm^{-1}$,Table S1: The sizes of blood components, based on data from Handin et al., Table S2: The RSD of the selected intensities of SERS signals of leucocytes, HeLa, and PC3 cells recorded from 1000 different spots within the same sample.

**Author Contributions:** Conceptualization, A.K. and T.S.; Methodology, K.N.; Software, K.N.; Validation, J.T.-D. and A.G.; Investigation, E.W., E.K.-G., W.Ś.; Data Curation, K.N.; Writing—Original Draft Preparation, A.K.; Writing—Review & Editing, A.K., T.S.; Funding Acquisition, A.K.

**Funding:** This research was funded by National Science Centre under grant UMO 2017/25/B/ST4/01109.

**Conflicts of Interest:** The authors declare no conflicts of interest.

## References

1. Ghossein, R.A.; Bhattacharya, S.; Rosai, J. Molecular Detection of Micrometastases and Circulating Tumor Cells in Solid Tumors. *Clin. Cancer Res.* **1999**, *5*, 1950–1960. [PubMed]
2. Swaby, R.F.; Cristofanilli, M. Circulating tumor cells in breast cancer: A tool whose time has come of age. *BMC Med.* **2011**, *9*, 43. [CrossRef] [PubMed]
3. Danila, D.C.; Fleisher, M.; Scher, H.I. Circulating Tumor Cells as Biomarkers in Prostate Cancer. *Clin. Cancer Res.* **2011**, *17*, 3903–3912. [CrossRef] [PubMed]
4. Tanaka, F.; Yoneda, K.; Kondo, N.; Hashimoto, M.; Takuwa, T.; Matsumoto, S.; Okumura, Y.; Rahman, S.; Tsubota, N.; Tsujimura, T.; et al. Circulating Tumor Cell as a Diagnostic Marker in Primary Lung Cancer. *Clin. Cancer Res.* **2009**, *15*, 6980–6986. [CrossRef] [PubMed]
5. Alix-Panabieres, C.; Pantel, K. Challenges in circulating tumour cell research. *Nat. Rev. Cancer* **2014**, *14*, 623–631. [CrossRef] [PubMed]
6. Nagrath, S.; Sequist, L.V.; Maheswaran, S.; Bell, D.W.; Irimia, D.; Ulkus, L.; Smith, M.R.; Kwak, E.L.; Digumarthy, S.; Muzikansky, A.; et al. Isolation of rare circulating tumour cells in cancer patients by microchip technology. *Nature* **2007**, *450*, 1235–1239. [CrossRef] [PubMed]
7. Andergassen, U.; Kolbl, A.C.; Mahner, S.; Jeschke, U. Real-time RT-PCR systems for CTC detection from blood samples of breast cancer and gynaecological tumour patients (Review). *Oncol. Rep.* **2016**, *35*, 1905–1915. [CrossRef] [PubMed]
8. Stott, S.L.; Hsu, C.H.; Tsukrov, D.I.; Yu, M.; Miyamoto, D.T.; Waltman, B.A.; Rothenberg, S.M.; Shah, A.M.; Smas, M.E.; Korir, G.K.; et al. Isolation of circulating tumor cells using a microvortex-generating herringbone-chip. *Proc. Natl. Acad. Sci. USA* **2010**, *107*, 18392–18397. [CrossRef] [PubMed]
9. Gleghorn, J.P.; Pratt, E.D.; Denning, D.; Liu, H.; Bander, N.H.; Tagawa, S.T.; Nanus, D.M.; Giannakakou, P.A.; Kirby, B.J. Capture of circulating tumor cells from whole blood of prostate cancer patients using geometrically enhanced differential immunocapture (GEDI) and a prostate-specific antibody. *Lab Chip* **2010**, *10*, 27–29. [CrossRef] [PubMed]
10. Ozkumur, E.; Shah, A.M.; Ciciliano, J.C.; Emmink, B.L.; Miyamoto, D.T.; Brachtel, E.; Yu, M.; Chen, P.-i.; Morgan, B.; Trautwein, J.; et al. Inertial Focusing for Tumor Antigen–Dependent and –Independent Sorting of Rare Circulating Tumor Cells. *Sci. Transl. Med.* **2013**, *5*, ra47–ra179. [CrossRef] [PubMed]
11. Krebs, M.G.; Hou, J.-M.; Ward, T.H.; Blackhall, F.H.; Dive, C. Circulating tumour cells: Their utility in cancer management and predicting outcomes. *Ther. Adv. Med. Oncol.* **2010**, *2*, 351–365. [CrossRef] [PubMed]
12. Yu, M.; Stott, S.; Toner, M.; Maheswaran, S.; Haber, D.A. Circulating tumor cells: Approaches to isolation and characterization. *J. Cell Biol.* **2011**, *192*, 373–382. [CrossRef] [PubMed]
13. Pantel, K.; Alix-Panabieres, C. Real-time liquid biopsy in cancer patients: Fact or fiction? *Cancer Res.* **2013**, *73*, 6384–6388. [CrossRef] [PubMed]
14. Gorges, T.M.; Pantel, K. Circulating tumor cells as therapy-related biomarkers in cancer patients. *Cancer Immunol. Immunother. CII* **2013**, *62*, 931–939. [CrossRef] [PubMed]

15. Jin, C.; McFaul, S.M.; Duffy, S.P.; Deng, X.; Tavassoli, P.; Black, P.C.; Ma, H. Technologies for label-free separation of circulating tumor cells: From historical foundations to recent developments. *Lab Chip* **2014**, *14*, 32–44. [CrossRef] [PubMed]
16. Yoon, H.J.; Kozminsky, M.; Nagrath, S. Emerging role of nanomaterials in circulating tumor cell isolation and analysis. *ACS Nano* **2014**, *8*, 1995–2017. [CrossRef] [PubMed]
17. Whitesides, G.M. The origins and the future of microfluidics. *Nature* **2006**, *442*, 368. [CrossRef] [PubMed]
18. Cristofanilli, M.; Budd, G.T.; Ellis, M.J.; Stopeck, A.; Matera, J.; Miller, M.C.; Reuben, J.M.; Doyle, G.V.; Allard, W.J.; Terstappen, L.W.M.M.; et al. Circulating Tumor Cells, Disease Progression, and Survival in Metastatic Breast Cancer. *N. Engl. J. Med.* **2004**, *351*, 781–791. [CrossRef] [PubMed]
19. Olmos, D.; Arkenau, H.T.; Ang, J.E.; Ledaki, I.; Attard, G.; Carden, C.P.; Reid, A.H.; A'Hern, R.; Fong, P.C.; Oomen, N.B.; et al. Circulating tumour cell (CTC) counts as intermediate end points in castration-resistant prostate cancer (CRPC): A single-centre experience. *Ann. Oncol.* **2009**, *20*, 27–33. [CrossRef] [PubMed]
20. Cohen, S.J.; Punt, C.J.; Iannotti, N.; Saidman, B.H.; Sabbath, K.D.; Gabrail, N.Y.; Picus, J.; Morse, M.A.; Mitchell, E.; Miller, M.C.; et al. Prognostic significance of circulating tumor cells in patients with metastatic colorectal cancer. *Ann. Oncol.* **2009**, *20*, 1223–1229. [CrossRef] [PubMed]
21. Allard, W.J.; Matera, J.; Miller, M.C.; Repollet, M.; Connelly, M.C.; Rao, C.; Tibbe, A.G.; Uhr, J.W.; Terstappen, L.W. Tumor cells circulate in the peripheral blood of all major carcinomas but not in healthy subjects or patients with nonmalignant diseases. *Clin. Cancer Res.* **2004**, *10*, 6897–6904. [CrossRef] [PubMed]
22. Neugebauer, U.; Bocklitz, T.; Clement, J.H.; Krafft, C.; Popp, J. Towards detection and identification of circulating tumour cells using Raman spectroscopy. *Analyst* **2010**, *135*, 3178–3182. [CrossRef] [PubMed]
23. Campion, A.; Kambhampati, P. Surface-enhanced Raman scattering. *Chem. Soc. Rev.* **1998**, *27*, 241–250. [CrossRef]
24. Ru, E.C.L.; Etchegoin, P.G. Single-Molecule Surface-Enhanced Raman Spectroscopy. *Annu. Rev. Phys. Chem.* **2012**, *63*, 65–87. [PubMed]
25. Maiti, K.K.; Samanta, A.; Vendrell, M.; Soh, K.S.; Olivo, M.; Chang, Y.T. Multiplex cancer cell detection by SERS nanotags with cyanine and triphenylmethine Raman reporters. *Chem. Commun.* **2011**, *47*, 3514–3516. [CrossRef] [PubMed]
26. Nima, Z.A.; Mahmood, M.; Xu, Y.; Mustafa, T.; Watanabe, F.; Nedosekin, D.A.; Juratli, M.A.; Fahmi, T.; Galanzha, E.I.; Nolan, J.P.; et al. Circulating tumor cell identification by functionalized silver-gold nanorods with multicolor, super-enhanced SERS and photothermal resonances. *Sci. Rep.* **2014**, *4*, 4752. [CrossRef] [PubMed]
27. Hoonejani, M.R.; Pallaoro, A.; Braun, G.B.; Moskovits, M.; Meinhart, C.D. Quantitative multiplexed simulated-cell identification by SERS in microfluidic devices. *Nanoscale* **2015**, *7*, 16834–16840. [CrossRef] [PubMed]
28. Zhang, P.; Zhang, R.; Gao, M.; Zhang, X. Novel Nitrocellulose Membrane Substrate for Efficient Analysis of Circulating Tumor Cells Coupled with Surface-Enhanced Raman Scattering Imaging. *ACS Appl. Mater. Interfaces* **2014**, *6*, 370–376. [CrossRef] [PubMed]
29. Jun, B.H.; Noh, M.S.; Kim, J.; Kim, G.; Kang, H.; Kim, M.S.; Seo, Y.T.; Baek, J.; Kim, J.H.; Park, J.; et al. Multifunctional Silver-Embedded Magnetic Nanoparticles as SERS Nanoprobes and Their Applications. *Small* **2010**, *6*, 119–125. [CrossRef] [PubMed]
30. Wang, X.; Qian, X.M.; Beitler, J.J.; Chen, Z.G.; Khuri, F.R.; Lewis, M.M.; Shin, H.J.C.; Nie, S.M.; Shin, D.M. Detection of Circulating Tumor Cells in Human Peripheral Blood Using Surface-Enhanced Raman Scattering Nanoparticles. *Cancer Res.* **2011**, *71*, 1526–1532. [CrossRef] [PubMed]
31. Chang, Z.M.; Wang, Z.; Shao, D.; Yue, J.; Xing, H.; Li, L.; Ge, M.; Li, M.; Yan, H.; Hu, H.; et al. Shape Engineering Boosts Magnetic Mesoporous Silica Nanoparticle-Based Isolation and Detection of Circulating Tumor Cells. *ACS Appl. Mater. Interfaces* **2018**, *10*, 10656–10663. [CrossRef] [PubMed]
32. Shi, W.; Paproski, R.J.; Moore, R.B.; Zemp, R. *Detection of Circulating Tumor Cells Using Targeted Surface-Enhanced Raman Scattering Nanoparticles and Magnetic Enrichment*; SPIE: Bellingham, WA, USA, 2014; p. 8.
33. Krafft, C.; Beleites, C.; Schie, I.W.; Clement, J.H.; Popp, J. Raman-based Identification of Circulating Tumor Cells for Cancer Diagnosis. In *Biomedical Vibrational Spectroscopy 2016: Advances in Research and Industry*; Mahadevan-Jansen, A., Petrich, W., Eds.; Spie-Int Soc Optical Engineering: Bellingham, WA, USA, 2016; Volume 9704.

34. Wang, H.H.; Liu, C.Y.; Wu, S.B.; Liu, N.W.; Peng, C.Y.; Chan, T.H.; Hsu, C.F.; Wang, J.K.; Wang, Y.L. Highly Raman-Enhancing Substrates Based on Silver Nanoparticle Arrays with Tunable Sub-10 nm Gaps. *Adv. Mater.* **2006**, *18*, 491–495. [CrossRef]
35. Dick, L.A.; McFarland, A.D.; Haynes, C.L.; Van Duyne, R.P. Metal Film over Nanosphere (MFON) Electrodes for Surface-Enhanced Raman Spectroscopy (SERS): Improvements in Surface Nanostructure Stability and Suppression of Irreversible Loss. *J. Phys. Chem. B* **2002**, *106*, 853–860. [CrossRef]
36. Fromm, D.P.; Sundaramurthy, A.; Kinkhabwala, A.; Schuck, P.J.; Kino, G.S.; Moerner, W.E. Exploring the chemical enhancement for surface-enhanced Raman scattering with Au bowtie nanoantennas. *J. Chem. Phys.* **2006**, *124*, 61101. [CrossRef] [PubMed]
37. Alvarez-Puebla, R.; Cui, B.; Bravo-Vasquez, J.-P.; Veres, T.; Fenniri, H. Nanoimprinted SERS-Active Substrates with Tunable Surface Plasmon Resonances. *J. Phys. Chem. C* **2007**, *111*, 6720–6723. [CrossRef]
38. Semin, D.J.; Rowlen, K.L. Influence of vapor deposition parameters on SERS active Ag film morphology and optical properties. *Anal. Chem.* **1994**, *66*, 4324–4331. [CrossRef]
39. Faulds, K.; Smith, W.E.; Graham, D. Evaluation of Surface-Enhanced Resonance Raman Scattering for Quantitative DNA Analysis. *Anal. Chem.* **2004**, *76*, 412–417. [CrossRef] [PubMed]
40. Szymborski, T.; Witkowska, E.; Adamkiewicz, W.; Waluk, J.; Kamińska, A. Electrospun polymer mat as a SERS platform for the immobilization and detection of bacteria from fluids. *Analyst* **2014**, *139*, 5061–5064. [CrossRef] [PubMed]
41. Witkowska, E.; Szymborski, T.; Kamińska, A.; Waluk, J. Polymer mat prepared via Forcespinning™ as a SERS platform for immobilization and detection of bacteria from blood plasma. *Mater. Sci. Eng. C* **2017**, *71*, 345–350. [CrossRef] [PubMed]
42. Kijeńska, E.; Swieszkowski, W. 2–General requirements of electrospun materials for tissue engineering: Setups and strategy for successful electrospinning in laboratory and industry. In *Electrospun Materials for Tissue Engineering and Biomedical Applications*; Uyar, T., Kny, E., Eds.; Woodhead Publishing: Sawston, UK, 2017; pp. 43–56.
43. Zhang, F.; Chen, P.; Zhang, L.; Mao, S.C.; Lin, L.; Tang, Y.G.; Cui, J.C.; Qi, X.D.; Yang, J.H.; Ma, Y.F. Enhancement of Raman scattering by field superposition of rough submicrometer silver particles. *Appl. Phys. Lett.* **2012**, *100*, 4. [CrossRef]
44. Zheng, S.; Lin, H.K.; Lu, B.; Williams, A.; Datar, R.; Cote, R.J.; Tai, Y.C. 3D microfilter device for viable circulating tumor cell (CTC) enrichment from blood. *Biomed. Microdevices* **2011**, *13*, 203–213. [CrossRef] [PubMed]
45. Cheng, W.T.; Liu, M.T.; Liu, H.N.; Lin, S.Y. Micro-Raman spectroscopy used to identify and grade human skin pilomatrixoma. *Microsc. Res. Tech.* **2005**, *68*, 75–79. [CrossRef] [PubMed]
46. Neugebauer, U.; Clement, J.H.; Bocklitz, T.; Krafft, C.; Popp, J. Identification and differentiation of single cells from peripheral blood by Raman spectroscopic imaging. *J. Biophotonics* **2010**, *3*, 579–587. [CrossRef] [PubMed]
47. Stone, N.; Kendall, C.; Smith, J.; Crow, P.; Barr, H. Raman spectroscopy for identification of epithelial cancers. *Faraday Discuss.* **2004**, *126*, 141–157. [CrossRef] [PubMed]
48. Notingher, I.; Green, C.; Dyer, C.; Perkins, E.; Hopkins, N.; Lindsay, C.; Hench, L.L. Discrimination between ricin and sulphur mustard toxicity in vitro using Raman spectroscopy. *J. R. Soc. Interface* **2004**, *1*, 79–90. [CrossRef] [PubMed]
49. Stone, N.; Kendall, C.; Shepherd, N.; Crow, P.; Barr, H. Near-infrared Raman spectroscopy for the classification of epithelial pre-cancers and cancers. *J. Raman Spectrosc.* **2002**, *33*, 564–573. [CrossRef]
50. Fung, M.F.K.; Senterman, M.K.; Mikhael, N.Z.; Lacelle, S.; Wong, P.T.T. Pressure-tuning Fourier transform infrared spectroscopic study of carcinogenesis in human endometrium. *Biospectroscopy* **1996**, *2*, 155–165. [CrossRef]
51. Huang, Z.; McWilliams, A.; Lui, H.; McLean, D.I.; Lam, S.; Zeng, H. Near-infrared Raman spectroscopy for optical diagnosis of lung cancer. *Int. J. Cancer* **2003**, *107*, 1047–1052. [CrossRef] [PubMed]
52. Utzinger, U.; Heintzelman, D.L.; Mahadevan-Jansen, A.; Malpica, A.; Follen, M.; Richards-Kortum, R. Near-Infrared Raman Spectroscopy for in Vivo Detection of Cervical Precancers. *Appl. Spectrosc.* **2001**, *55*, 955–959. [CrossRef]
53. Krafft, C.; Knetschke, T.; Siegner, A.; Funk, R.H.W.; Salzer, R. Mapping of single cells by near infrared Raman microspectroscopy. *Vib. Spectrosc.* **2003**, *32*, 75–83. [CrossRef]

54. Taleb, A.; Diamond, J.; McGarvey, J.J.; Beattie, J.R.; Toland, C.; Hamilton, P.W. Raman Microscopy for the Chemometric Analysis of Tumor Cells. *J. Phys. Chem. B* **2006**, *110*, 19625–19631. [CrossRef] [PubMed]
55. Nijssen, A.; Bakker Schut, T.C.; Heule, F.; Caspers, P.J.; Hayes, D.P.; Neumann, M.H.; Puppels, G.J. Discriminating basal cell carcinoma from its surrounding tissue by Raman spectroscopy. *J. Investig. Dermatol.* **2002**, *119*, 64–69. [PubMed]
56. Notingher, I.; Verrier, S.; Haque, S.; Polak, J.M.; Hench, L.L. Spectroscopic study of human lung epithelial cells (A549) in culture: Living cells versus dead cells. *Biopolymers* **2003**, *72*, 230–240. [CrossRef] [PubMed]
57. Uzunbajakava, N.; Lenferink, A.; Kraan, Y.; Willekens, B.; Vrensen, G.; Greve, J.; Otto, C. Nonresonant Raman imaging of protein distribution in single human cells. *Biopolymers* **2003**, *72*, 1–9. [CrossRef] [PubMed]
58. El-Said, W.A.; Kim, T.H.; Kim, H.; Choi, J.W. Analysis of intracellular state based on controlled 3D nanostructures mediated surface enhanced Raman scattering. *PLoS ONE* **2011**, *6*, e15836. [CrossRef] [PubMed]

nanomaterials

MDPI

Review

# Analysis of Biomolecules Based on the Surface Enhanced Raman Spectroscopy

**Min Jia [1,2], Shenmiao Li [3], Liguo Zang [1], Xiaonan Lu [3] and Hongyan Zhang [1,*]**

[1] Shandong Provincial Key Laboratory of Animal Resistance Biology, Institute of Biomedical Sciences, Key Laboratory of Food Nutrition and Safety of Shandong Normal University, College of Life Science, Shandong Normal University, Jinan 250014, China; mjia@sdnu.edu.cn (M.J.); zangliguo@163.com (L.Z.)

[2] Beijing Advanced Innovation Center for Food Nutrition and Human Health, Beijing Technology and Business University (BTBU), Beijing 100048, China

[3] Food, Nutrition and Health Program, Faculty of Land and Food Systems, The University of British Columbia, Vancouver, BC V6T 1Z4, Canada; shenmiao.ivy.li@gmail.com (S.L.); xiaonan.lu@ubc.ca (X.L.)

* Correspondence: zhanghongyan@sdnu.edu.cn; Tel.: +86-531-861-80745

Received: 28 August 2018; Accepted: 14 September 2018; Published: 15 September 2018

**Abstract:** Analyzing biomolecules is essential for disease diagnostics, food safety inspection, environmental monitoring and pharmaceutical development. Surface-enhanced Raman spectroscopy (SERS) is a powerful tool for detecting biomolecules due to its high sensitivity, rapidness and specificity in identifying molecular structures. This review focuses on the SERS analysis of biomolecules originated from humans, animals, plants and microorganisms, combined with nanomaterials as SERS substrates and nanotags. Recent advances in SERS detection of target molecules were summarized with different detection strategies including label-free and label-mediated types. This comprehensive and critical summary of SERS analysis of biomolecules might help researchers from different scientific backgrounds spark new ideas and proposals.

**Keywords:** biomolecule; surface enhanced Raman spectroscopy (SERS); nanomaterial; analysis; identification; biology

## 1. Introduction

The analysis of biomolecule is significant in various application fields, such as clinical diagnostics, food safety analysis, environmental monitoring and pharmaceutical development [1–4]. Biomolecules are the substances present in living organisms that play important roles in chemical and biological processes, including macromolecules (e.g., proteins, nucleic acids, carbohydrates, lipids) and small molecules (e.g., primary metabolites, natural products) [5–7]. Many detection methods have been employed to determine the biomolecules in vitro or in vivo, such as chromatography-mass spectroscopy [8], enzyme-linked immunosorbent assay (ELISA) [9], colorimetric and fluorescence detection [10,11], polymerase chain reaction (PCR) [12,13], and vibrational spectroscopy (e.g., infrared and Raman spectroscopy) [14]. This review focuses on the latest advances of surface enhanced Raman spectroscopy (SERS) in the analysis of biomolecules in humans, animals, plants and microorganisms over the past ten years (Figure 1).

SERS provides complementary analysis on molecular identification or quantification because it can reveal the information of complete structure or accurate amount [15]. Nanomaterials have been introduced into SERS methods as substrates and nanotags to improve the sensitivity and selectivity of the detection with excellent enhancement factor that can realize single target molecule detection [16]. With the development of nanotechnology, increasing varieties of nanomaterials were discovered, including noble metal (e.g., Au and Ag) and transition metal (e.g., Co and Pt) with different material morphologies, such as nanoparticles (NPs), nanoflowers, nanoclusters, and nanostars. Due to the synergistic effects, many nanohybrids were designed with non-metallic material, such as $SiO_2$ [17] and graphene oxide (GO) [18]. Numerous complex structures were designed to enhance the SERS signal, containing hexagonal-packed lotus seedpod like array substrate [19], $Fe_3O_4$-Au core-shell NPs with branched gold shell [20] and Si nanopillars (SiNPLs)@silver nanoparticles (AgNPs) [21], etc.

Based on the structures of analyte molecules, different SERS detection strategies, including label-free and label-mediated types, were developed to achieve sensitive and accurate analysis. Label-free strategies can achieve the direct detection of target molecules without any Raman labels [22]. At first, the SERS signals are directly generated due to the interaction between target molecules and SERS-active substrates. In recent years, to overcome the challenges of deciphering the complex SERS spectra and low sensitivity of the previous methods, other label-free methods, such as aggregation, were developed to realize a more sensitive detection. Analyte-induced aggregation of plasmonic NPs leaded to the formation of hot spots that can enhanced the Raman scattering signals because of the particle size increase [23]. Due to the difficulty in direct detection of biomolecules, label-mediated detections were applied to specific detection by the use of antibodies or aptamers [24,25]. The general label-mediated strategies were immunoassay-based strategies adopting antigen-antibody reaction and SERS nanotags in SERS analysis to improve the sensitivity and selectivity of target detection. SERS nanotags were prepared by the combination of Raman reporter, metallic NPs and target recognition elements. Besides the general antibodies, aptamers were also used as the recognition elements. With the high specific biorecognition of aptamer and its unique DNA feather such as the rolling circle amplification technique, aptamer-based strategy for constructing SERS biosensors has attracted increasing attention in biomolecule detection.

In this review, we highlighted the application of SERS detection of biomolecules that are body components of human, animal, plant and microorganism. The SERS detection methods were summarized and classified based on the natural property of target molecules.

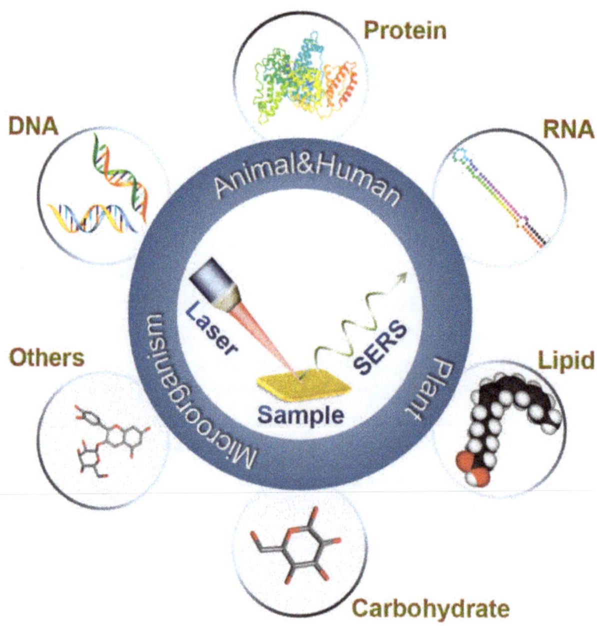


**Figure 1.** SERS detection of biomolecules.

## 2. Application on the Detection of Human and Animal Original Biomolecule

### *2.1. Protein*

The accurate and sensitive detection of proteins is crucial for proteomics and therapeutic research. Proteins are essential parts of organisms that are involved in virtual process with various functions. Therefore, structure analysis and quantitation of major species of functional proteins are critically important for understanding their functions. Meanwhile, there is an urgent demand for rapid and sensitive analysis of disease biomarkers for the early diagnosis of diseases and raising the survival rate of patients.

#### 2.1.1. Crucial Functional Proteins

Albumins generally act as transport proteins for numerous compounds in blood plasma [26]. Bovine serum albumin (BSA) was chosen as a model molecule to research the enhancement of SERS-active substrates in label-free detection. Noble metallic nanomaterials such as gold nanocylinders [22] and $SiO_2$/Au nanoshells [17] were used as substrates for BSA detection. To keep proteins in their natural structure and conformation, in-liquid SERS detection of protein was based on nanoparticle aggregating as SERS-active hot spots for label-free detection [27]. Foti et al. [26] applied optically induced gold nanorods aggregation to the detection of BSA in liquid by combining light scattering, plasmon resonance and SERS. Other Raman resonant biomolecules such as catalase and hemoglobin were used to investigate the application of this methodology, and the limits of detection of catalase and hemoglobin were much lower than that of BSA.

Hemoglobins are a kind of iron-containing metalloproteins that transport oxygen in erythrocytes of all vertebrates. Iron porphyrins in hemoglobins can be an intrinsic selective probing which shows resonance enhanced Raman spectrum when excited with a laser in Soret and Q-band regions [15]. Therefore, label-free detection strategy could be applied to hemoglobin detection. Silver nanomaterials, such as colloidal AgNPs [15] and nano-silver film [28], were usually used as substrates in SERS experiment to achieve excellent enhancement of Raman signal. A novel SERS nanoprobe modified with organic cyanide (4-mercaptobenzonitrile, MBN) was developed for accurate detection of oxidized hemoglobin and $Fe^{3+}$ ions in living cell [7]. The MBN-modified, SERS-active nanopipette gave new insights into in situ single cell detection.

Enzymes, which are mostly proteins in nature and act as selective biological catalysts, were characterized by SERS. Thrombin is one kind of serine proteolytic enzyme that plays an important role in regulating blood coagulation. Due to the successful selection of the thrombin-binding aptamer with high affinity and high specificity, a lot of aptamer-based SERS methods were developed for tracing thrombin analysis with improved selectivity and sensitivity [25,29]. The thrombin-binding aptamer was used as the capture probe with selective affinity that realized the SERS application in complex matrices. Meanwhile, by using the unique catalytic activity of thrombin, the enzymatic amplification-based strategy was established to achieve sensitive and selective detection of thrombin. The addition of thrombin could catalytically cleave the multiple arginine peptides into fragments, so that the aggregates of gold nanoparticles (AuNPs) incorporating Raman reporter molecules induced by peptides was decreased [30]. As the ability of fragments to form "hot spots" being weakened, the SERS signals were sharply diminished. The limit of detection (LOD) was 160 fM, which indicated that the sensitivity of this method was improved compared with non-enzymatic amplification based methods. This strategy can also be applied to other enzymes by appropriately utilizing the catalytic properties. Kinases are a group of enzymes that catalyze phosphorylation of specific substrate proteins [31–33]. Using this property, the kinase was detected by monitoring the changes in the intensity of Raman peak markers before and after phosphorylation [34]. The quantification of activity level of the enzyme can be achieved utilizing the enzyme concentration based on the enzymatic amplification-based strategy [35].

Immunoglobulin, also known as the antibody, is Y-shaped protein in different varieties known as isotypes [36,37]. Immunoglobulin is generated by the immune system to counteract pathogens such as viruses [38] and bacteria [39]. Based on the inherent antigen-antibody reaction, the immunoassay-based strategy was employed to detect immunoglobulin [40]. The traditional sandwich-format assays were usually achieved by combining antibody modified SERS-active substrates to capture target immunoglobulins and extrinsic Raman labels to enhance antigen-antibody binding kinetics [41]. Neng et al. [42] utilized AuNPs coated with the antigen as the SERS-active substrate and protein A/G modified with Raman reporter molecules as a bi-functional reporter to detect the antibody of West Nile Virus. The LOD was 2 ng/mL antibodies in the serum sample. Variants of this detection platform were applied to the trace-level detection of immunoglobulin.

Cytokines are secreted proteins produced by immune system cells regulating the immune activity [43,44]. Immunoassay-based strategy was also applicable to cytokines detection. Kamińska et al. [45] developed an immunoassay-based strategy employing diatom biosilica as the SERS-active substrate and DTNB-labeled immune-AuNPs as the SERS nanotag to detect the interleukin 8 (IL-8) in blood plasma [46]. Based on this method, a SERS immunoassay combined with a microfluidic device was developed for multiplexed recognition of interleukins IL-6, IL-8, and IL-18 in blood plasma with the LOD of 4.2 pg/mL. Wang et al. [47] also developed a multiplexed immunoassay of three cytokines, interferon gamma, interleukin-2, and tumor necrosis factor alpha, based on SERS signal enhanced by the controlled assembly of "hot spot" with low LOD and large signal-to-noise ratio in complex matrix.

Hormones communicate among organs and tissues to regulate various functions as a class of signaling molecules [48]. Human chorionic gonadotropin (hCG), an important pregnancy diagnostic marker, is a glycoprotein hormone that stimulates steroid hormone and progesterone production in the luteum [49]. Liang et al. [18] developed a series of methods based on the nanogold reaction between $HAuCl_4$ and $H_2O_2$ that forms AuNPs with Victoria blue 4R molecular probes to detect hCG. The concentration of hCG can influence the catalytic effect of nanozymes such as AgNPs clusters and GO. Therefore, the changes of SERS intensity were linear to the concentration of hCG. This novel strategy can be further applied to other proteins combined with the nanozyme catalysis to develop the SERS detection platform.

Receptors are protein molecules that bound with extracellular chemical signals, causing some forms of cellular/tissue responses [50]. G-protein-coupled receptor 120 (GPR120) mediated response to long chain fatty acids (LCFAs), but the mechanism of GPR120 acting in the transduction of LCFAs was uncovered. To understand the function of GPR120 in fat chemoreception, SERS-active gold nanorods conjugated with fluorescence-active $CaMoO_4:Eu^{3+}$ NPs and Raman reporter molecule 4-mercaptobenzoic acid were used to achieve SERS-fluorescence imaging of GPR120 in a single cell [51].

### 2.1.2. Disease Biomarkers

Disease biomarker refers to an extracellular indicator of disease biological state or condition. These traceable substances in the organism indicate organ functions or other aspects of health. The quantitative and sensitive detection of biomarkers is very meaningful for early clinical diagnosis and evaluating the therapeutic response. Early diagnosis has great potential in improving the survival of patients with serious diseases, such as cancer, neurodegenerative disorders, and cardia-cerebrovascular diseases. SERS-based assays have attracted significant attention as a highly sensitive and non-destructive analysis with great potential in biomarker detections.

Cancer is a kind of malignant disease curable only in the early stage. Cancer biomarkers, typical group of proteins, are usually over-expressed during tumor progression [16]. Therefore, tumor biomarkers have already been used for specific cancer identification, including carcinoembryonic antigen (CEA) for colorectal cancer [24], $\alpha$-fetoprotein (AFP) for liver cancer [16] and prostate specific antigen (PSA) [52]. "Sandwich" structure SERS immunoassays were usually used in biomarker protein detection and involved SERS-active substrates, target proteins, and SERS nanotags [53]. The multiple

SERS-active substrates were prepared by conjugating antibodies to capture target proteins [3,24,54]. Then the nanostructure-based SERS nanotags modified with Raman reporter and target proteins participated the antibody-antigen-antibody interaction. Li et al. [24] utilized gold nanobowl arrays and gold nanoshells as SERS-active substrates and SERS nanotags, respectively. This sandwich immunoassay showed a good linear relationship between CEA concentration and SERS signal intensity with a LOD of 1.73 pg/mL CEAs. PSA was detected on a glass slide with spot-arrays using SERS dot as the probe (Figure 2A). A full-area confocal raster Raman was applied to detect the SERS dot at the single probe level fixed on the glass slide by antigen-antibody interaction. This method showed high sensitivity with a LOD of 3.4 fM PSAs [52]. Besides single target detection, the SERS-based multiplex immunoassays were utilized to the ultrasensitive detection of different cancer biomarkers simultaneously. Wang et al. [16] used two kinds of Raman reporters labeled $SiO_2$@Ag immune probes and gold-film hemisphere array substrate to realize multiple detection of PSA and AFP (Figure 2B). Gao et al. [55] developed a droplet-based microfluidic SERS sensor based on wash-free magnetic immunoassay technique. Using the microfluidic system integrated compartments for mixing, PSA in serum can be split and detected without any washing steps. Quantitative evaluation of PSA was realized using the Raman signals of the residual SERS nanotags in the large droplet.

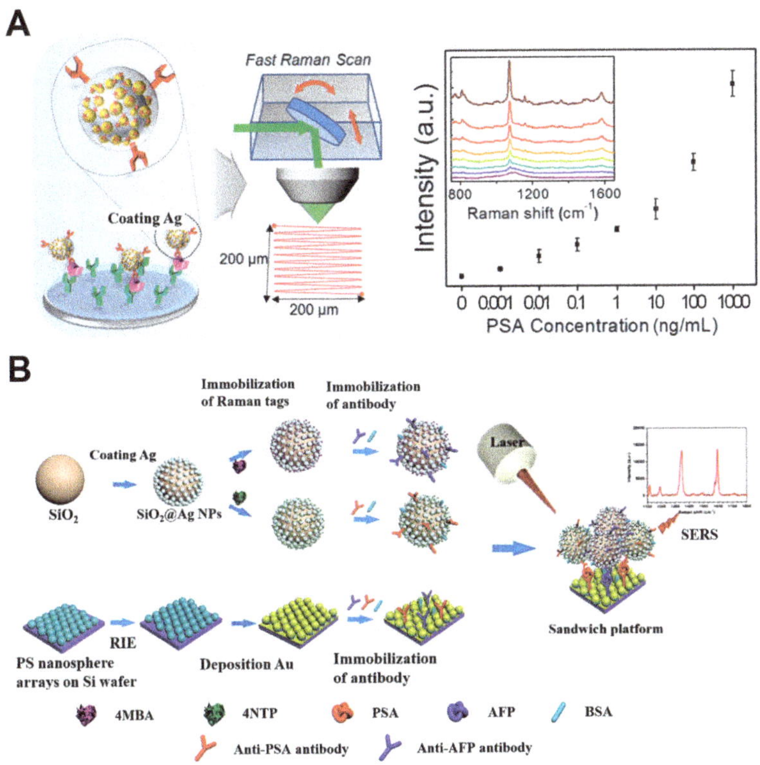


**Figure 2.** (**A**) Schematic of a SRES-based immunoassay for PSA detection [52]. (**B**) Schematic of a SERS-based multiplex immunoassay detection for PSA and AFP [16]. Reproduced with permission from [52]. Copyright American Chemical Society, 2016. Reproduced with permission from [16]. Copyright Elsevier, 2018.

C-reactive protein (CRP) has been widely studied to reveal the relevance of inflammation and cardiovascular diseases. Guo et al. [56] developed an enzymatic activation strategy combined with immunoassay analysis to activate reduction caged Raman reporters for SERS detection of CRP. Agarose beads modified with capture antibodies, CRP and horseradish peroxidase (HRP)-modified detection antibodies were used to constitute the sandwich assay. The HRP activated reduction caged LMG to generate SERS active MG for measurements. A linear curve on a logarithmic scale was obtained between CRP concentrations and SERS signals. In addition to the immunoassay strategy, label-free strategy was also suitable for CRP detection [57]. 3D AgNPs aggregates functionalized by phosphocholine were constructed to selectively capture CRP and to enhance the SERS efficiency accordingly.

Alzheimer's disease was a chronic and progressive neurodegenerative disease causing dementia in the senior population. Amyloid-β peptide, a biomarker to diagnose Alzheimer's disease, was hydrophobic in nature. Distinct from the above disease biomarkers, label-free strategy was usually used in SERS detection of amyloid-β. Due to synergistic effects, many nanohybrids were designed as SERS-active substrates. For amyloid-β detection, hexagonal-packed lotus seedpod like array substrate [19], biomimetic lipid membrane [58] and SiNPLs@AgNPs [21] were constructed. 3D SERS platforms [59] were also designed to enhance the "hot spot" formation in 3D. A 3D GO based SERS substrate decorated by core-shell nanoparticle was developed from hybrid 2D GO cross-linked by amine-modified PEG. The SERS enhancement factor for 3D SERS substrate was around $3.9 \times 10^{12}$ with the LOD of 500 fg/mL amyloid-β.

Neuron-specific enolase (NSE) is a mortality predictor in traumatic brain injury patients. "Sandwich" immunoassay methods were generally used in SERS detection of NSE with different substrates. Wang et al. [60] utilized an indium tin oxide conductive glass slip with hollow gold nanospheres (HAuNPs) as the SERS substrates instead of normal silver nanomaterials. Nile blue A and NSE antibody functionalized HAuNPs were used as SERS nanotags to achieve the sensitive detection of NSE with the LOD of 0.1 ng/mL. Subsequently, lateral flow glass-hemostix (FGH) combined with Au nanocage was used as SERS substrates to detect NSE in blood plasma achieving a LOD of 0.74 ng/mL [61]. To realize rapid and low-cost clinical diagnosis, a paper-based SERS lateral flow strip was developed to detect NSE. Au nanostar@Raman Reporter@silica NPs was employed as SERS nanotag that exhibits superior performance compared to the colorimetric methods with the LOD of 0.86 ng/mL [62], indicating that paper-based SERS detection had a great potential to meet the requirement of point-of-care (POC) testing.

### 2.2. *Nucleic Acid*

Detection of nucleic acids such as the DNA and RNA is significant for disease diagnosis and gene therapy [63–65]. SERS detection was widely used in nucleic acid analysis offering simple procedures, ultrasensitive and unique structural characterization. There are two major strategies for nucleic acid detection. Label-free detection strategy based on the distinctive SERS signals from four bases of DNA/RNA. The other strategy was modifying nucleic acid with extrinsic Raman labels or constructing SERS nanotags, which have more extensive applications than label-free strategy. A series of ingenious methods based on the properties of complementarity and amplification were developed for nucleic acid detection, such as the classical "sandwich" method, the hairpin DNA-assisted method and the signal amplification method.

#### 2.2.1. DNA

For label-free detection, the SERS signal was ordinarily generated by DNA directly absorbed on the surface of the SERS substrate. DNA bases showed specific Raman spectral which can be enhanced by the SERS-active substrate and distinctive with other impurities [66]. The facile laser scribing method was developed to fabricate AgNPs@GO composite film with a microfluidic chip for DNA detection [67]. Moreover, a unique label-free detection based on hairpin DNA and NPs in situ growth strategy was used in SERS biosensor [68]. Qian et al. utilized a peptide nucleic acid (PNA) in hairpin structure immobilized on a glass slide to recognize target DNA which can hybridize with PNA probe (Figure 3A). The duplex structure could adsorb positively charged silver ions that were chemically reduced to form AgNPs. The bases and AgNPs complex could yield a sensitive Raman signal which showed a good linear relationship with the DNA concentration. For the classical "sandwich" method, various SERS-active substrates and SERS nanotags were modified with the complementary strand to form target DNA-bridged sandwich complex. Yu et al. [69] employed two probe DNA-immobilized particles to specifically recognize the nucleotide binding sites of 683 and 735 positions on target prostate cancer antigen 3 (PCA3) mimic DNA separately via hybridization reactions. This method showed high sensitivity with the LOD of 2.7 fM PCA3. Fu et al. [70] developed a paper-based SERS lateral flow strip for sensitive detection of human immunodeficiency virus type 1 (HIV-1) DNA with Raman reporter modified AuNPs as the SERS nanotags. This method showed potential feasibility in POC self-diagnostics with LOD of 0.24 pg/mL. To improve the sensitivity of the biosensors, SERS nanotag was extensively used in DNA amplification method. Ye et al. [71] combined a triple-helix molecular switch with cascade signal amplification to achieve ultrasensitive detection of p53 gene. This amplification strategy can achieve a LOD as low as 21 aM, demonstrating a higher sensitivity.

The nucleobases are essential parts of the DNA construction and are involved in numerous processes in biology. SERS analysis was also applied to the detection of nucleobases that showed distinctive SERS signals. A microfluidic device combined with label-free SERS measurements was used to detect adenine with silver colloids as SERS substrate [72]. The prominent SERS peak at 770 $cm^{-1}$ was caused by the ring-breathing band of adenine. The concentration of adenine showed a good linear relationship with SERS intensity quantified by the peak area of SERS peak at 770 $cm^{-1}$. On the other hand, single nucleotide polymorphism in mitochondrial DNA (16189T→C) can also be detected by SERS analysis utilizing the ion-mediated cascade amplification strategy [73]. Target DNA binding could successfully introduce AgNPs combining with the DNA ligase reaction. By detecting the dissolved $Ag^+$ from AgNPs, the LOD of targeted DNA was as low as $3.0 \times 10^{-5}$ fm/μL of adenine.

#### 2.2.2. RNA

Similar to SERS detection of DNA, the strategies used in DNA detection were also efficient in RNA detection. For label-free detection, a variety of SERS-active substrates were designed to improve the Raman enhancement factor, such as hollow Au nanoflowers substrates [74], silver nanorod array substrates [75] and SWNTs@AgNPs [76]. Zheng et al. [76] developed an approach of DNA-templated in situ growth of AgNPs on SWNTs for the sensitive detection of a potential cancer marker miRNA-21. The silica microbeads-conjugated double stranded DNA competitively bound to the target miRNA, forming liberated ssDNA that showed high affinity to SWNTs. The liberated ssDNA-SWNT complex acted as a "nanoscaffold" for $Ag^+$ to form SWNT@AgNPs upon reduction. The SWNT@AgNPs biosensor can sensitively quantify miRNA-21 with a detection limit of 5 pM. To improve the sensitivity of the SERS detection, the sandwich strategy was also employed in detecting miRNA. A novel type of $Fe_3O_4$-Au core-shell NPs with branched gold shell integrated SERS activity and superparamagnetism was designed to form the sandwich structure for the detection of miRNA-21 as shown in Figure 3B [20]. The LOD of miRNA-21 in serum was as low as 623 aM that indicated the branched nanostructures were also appropriate for SERS detection beside of smooth surfaces nanostructures. Apart from sandwich DNA nanostructures, Ye et al. [77] developed a series of signal amplification methods for miRNA-21, miRNA-141 and miRNA-203 determination. A dual functional DNA-linker-DNA probe was used for a symmetric signal amplification reaction to simultaneously analyze miRNAs.

The sensitive and simultaneous detection of multiple microRNAs can also be achieved by the application of SERS, which holds great promise for the early diagnosis of various diseases. Zhou et al. [78] utilized multiple DNA modified SERS nanotags and Ag-HMSs SERS substrate to simultaneously detect three Hepatic carcinoma related miRNAs based on the sandwich hybridization assay with a LOD of 10 fM. Shin et al. [79] developed a SERS detection combined with field-flow fractionation to sensitively detect multiple miRNAs. The target-specific polystyrene (PS) particles of three different diameters were utilized for target miRNA binding. By the polyadenylation reaction, a long tail composed of adenine was generated and the high complementariness to polythymine conjugated AuNPs led to SERS sensitivity enhancement. The three size-coded complexes were distinguished by field-flow fractionation and Raman signals obtained from the separated PS probes to measure three miRNAs simultaneously.

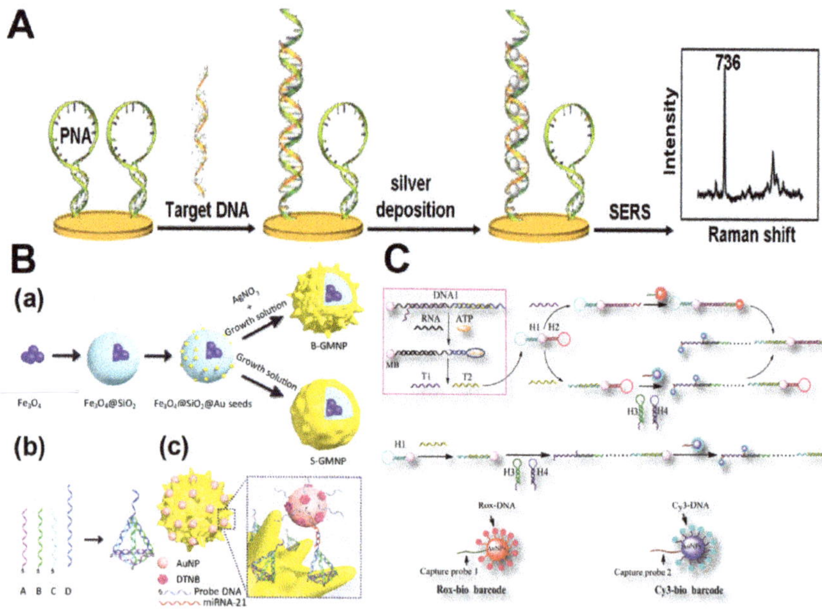


**Figure 3.** (**A**) Schematic of label-free strategy by using hairpin DNA [68]. (**B**) Sandwich strategy by using tetrahedral DNA (**a**) Schematic of B-GMNPs and S-GMNPs preparation; (**b**) tetrahedral DNA construction. (**c**) sandwich-structured strategy [20]. (**C**) Schematic of the asymmetric signal amplification SERS assay and process of HCR [77]. Reproduced with permission from [68]. Copyright Elsevier, 2018. Reproduced with permission from [20]. Copyright Springer, 2017. Reproduced with permission from [77]. Copyright American Chemical Society, 2015.

### 2.3. Carbohydrate

The measurement of glucose in biological systems is key to monitoring and regulating the blood glucose level for both diabetes patients and healthy individuals. To detect glucose at very low levels, the label-free strategy [80,81] and the enzyme-mediated catalytic strategy [82,83] were mainly utilized in SERS methods.

For the label-free detection of glucose, a series of SERS-active substrates were designed. Sooraj et al. [81] developed a coupled patterned Si with AgNPs arrays to improve the affinity of glucose with metal NPs. The limit of detection for glucose was low to $5 \times 10^{-5}$ g/mL, which was much lower than the blood glucose level. For enzyme-mediated catalytic strategy, glucose oxidase was utilized to catalyze glucose producing hydrogen peroxide ($H_2O_2$) to etch the AgNPs marked with 4-mercaptopyridine, leading to the decrease of the SERS signal. The lowest detectable concentration was 10 μM [82].

### 2.4. Others

Neurotransmitters are important for information transmission in the nervous system. Catecholamines are a kind of biological amine neurotransmitters consisting of amino and catechol in structure. The sensitive and accurate detection of catecholamines is significant for brain function research and neurological diseases monitoring. Dopamine is one of catecholamine class that plays an important role in human physiology. Tang et al. [84] developed an aptamer-induced Au@Ag nanorod dimer self-assembly method to detect the dopamine ultrasensitively with a LOD of 0.006 pM. Besides the detection methods based on aptamers and antibodies, Dumont et al. [85] developed a salt-induced colloid pre-aggregation strategy to overcome the protein corona stabilization for analysis of serum samples. Moreover, the analysis of multiple catecholamines in the complex was also achieved by SERS detection. Cao et al. [86] utilized an Au-Fe Raman label for the rapid detection of dopamine, norepinephrine or epinephrine in complex serum. Moody et al. [87] developed a SERS sensor for rapid analysis of seven neurotransmitters using AuNPs and AuNPs at multiple excitation wavelengths.

## 3. Application on the Detection of Plant Original Biomolecules

Compared to the application of SERS on the detection of animal original biomolecules, there were far fewer reports about the plant originated biomolecules, which are mainly concentrated in foods or medicinal plants, such as antioxidants, anthocyanins, Chinese herbal medicinal ingredients and so on. SERS also has been applied to the molecular fingerprint identification of plant while very few applications of SERS have been explored in the basis physiological study of model plants [88,89].

### 3.1. *Lipids and Antioxidant*

Lipids are the main food components, which are critical for nutrition concerns [90,91]. Study on their oxidative status can help the storage and process industry. Plants are known for their antioxidative effects due to their secondary metabolites, which can be used as food additives to avoid lipid oxidation. Unsaturated fatty acids also are important for the quality and bioactivity of lipids [92,93]. For the SERS detection of lipids and antioxidants, nanomaterials such as AgNPs and AuNPs were mainly used to fabricate the SERS-active substrate using the nanofabrication methods of *e*-beam lithography [94], electroless plating [95], and so on. Interaction of targets and nanomaterials was investigated in detail to improve the sensitivity of SERS detection.

SERS have favorable potential to offer exact information of the chemical constitution of the lipid and its oxidation state. Li et al. [89] investigated the SERS assay using silver (Ag) dendrites to enhance SERS signal of detecting canola oil and alpha-tocopherol, the oxidation process of canola oil was also investigated. SERS showed better sensitivity than general Raman or ultraviolet (UV) methods in monitoring the transformation of lipid structure when oxidated. Tomato represents a major horticultural crop in human diet. Carotenoids, a kind of antioxidant rich in tomatoes, are important for plant physiology and mammalian organisms [96]. Radu et al. studied the differences of two kinds of carotenoids in tomato extracts using an *e*-beam lithography SERS-active substrate. A model sample which was a mixture of two kinds of carotenoids was processed and analyzed. Two data processing methods were used, and the result agreed with the conventional analysis methods [94].

Chan et al. [97] described the preparation of nanoarrays with AgNPs used as SERS substrates. The nanoparticle-based platform offered application prospects in SERS analysis of beta-carotene, and the detection limit was <0.63 ppm. Hsueh et al. [95] created a facile designed 3D substrate for SERS detection of beta-carotene. The SERS devices had high-density, uniformly distributed hotspots of gold (Au) multibranches, which were electroless plated with a nanoporous polymer as a template. The sensitivity, stability and reproducibility were investigated. The new approach provided new substrate fabricating method for the analysis of analytes with an enhancement factor of $10^8$.

SERS was also used to detect butylated hydroxyanisole (BHA), generally used in foods or oil products to avoid oxidation [98]. Direct determination of BHA in several edible and essential oils without any sample treatment was realized by using SERS with a low-cost homemade silver substrate. The detailed conditions of experiment, characterics of every kind of oil, and the traits of substrate were showed and discussed [99]. Interaction of targets and nanomaterials is crucial for the improvement of the LOD of a SERS based detection. Studies on SERS detection of catechin showed that the ratio of catechin to AgNPs and NaCl was important for the data quality. Raman signal was detectable at $10^{-18}$ M level at the optimized ratio. Raman detection of catechin with high sensitivity and reproducibility was achieved due to the obvious enhancement of Raman signal [100]. Aguilar-Hernandez et al. [88] systematically evaluated SERS measurements of several examples of antioxidants compounds derived from various plants.

### 3.2. Anthocyanins

Anthocyanins are important natural products leading to the various colors in different parts of plants, including fruits, flowers, and grains [101,102]. Anthocyanins are derivatives of the salts of 2-phenylbenzopyrylium, naturally present as glycosylated molecules [103,104]. Subtle changes of the molecule structure can be revealed in the SERS spectra [105], which demonstrated the advantages of SERS used to identificate anthocyanidins. Benzopyrylium is the common moiety of the molecule structure of anthocyanidins with different phenyl rings, and different kinds of anthocyanidins exist in aqueous extracting solutions with different pH. Compared to resonance Raman, which could not distinguish the similar species, SERS can be used to further identify anthocyanidins. Zaffino et al. [106] identified some kinds of anthocyanidins using SERS and studied the influence of the pH on the SERS spectrum of six main anthocyanidins. They further provided SERS procedures of identificating anthocyanin in plant extracts. Optimized procedures of sample extraction and preparation were selected for different plant species and then detected by SERS measurement [107]. Luca et al. [108] reported the SERS spectra of analytes extracted from mulberry, gromwell, rhubarb and so on. It was proposed that the SERS spectrum was correlated with the molecule structure.

### 3.3. Chinese Herbal Medicinal Ingredients

SERS is also used to characterize herb extracts. Gu et al. [109] presented an analytical method for the rapid detection and identification of bioactive substances from Chinese herbs by combining thin layer chromatography (TLC) and SERS. The limits of detection were 0.05–0.10 μM, which were far more sensitive than the UV lamp based method. The established method further enabled predicting and uncovering of unknown substances from Chinese herbs. The TLC-SERS strategy for the sample preparation procedure of *n*-butanol was illustrated in Figure 4.

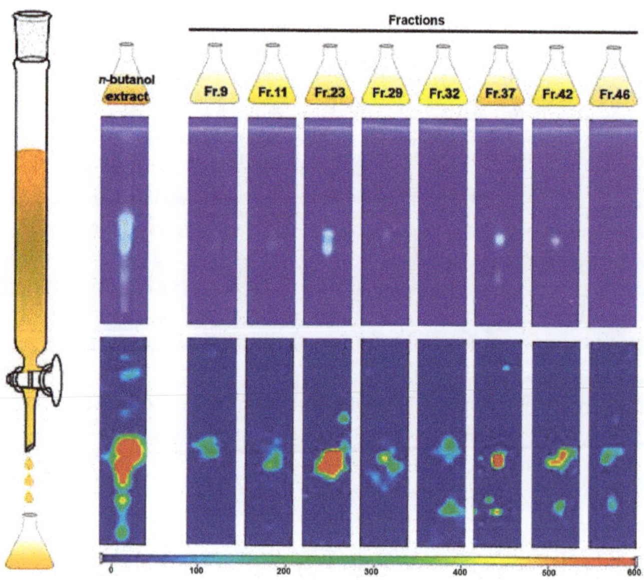


**Figure 4.** Schematic of the proposed TLC-SERS method for n-butanol extract detection [109]. Reproduced with permission from [109]. Copyright Elsevier, 2018.

Zhang et al. [110] employed SERS and fluorescence spectroscopy to test the interaction of herb molecules with human serum albumin (HSA). The SERS methods were applied to predicting the molecule conformation on colloidal AuNPs. Similar transformations were found for four ginsenosides when combined with HSA, while the glucose and aglycone were exposed to fit suitable sites. Zheng et al. [111] developed a novel SERS method to monitor 5-demethylnobiletin produced in citrus and the SERS methods based on substrate or solution were all well correlated with high performance liquid chromatography (HPLC). The solution-based SERS method separated nobiletin by applying a procedure like "affinity chromatography". The substrate-based SERS could simply and quickly collect the "fingerprint" spectra. SERS had more advantages than HPLC methods in convenient, rapid characterizition and quantification of the production of 5-demethylnobiletin.

### *3.4. Molecular Fingerprint Identification of Plant*

The fingerprinting method is widely applied to the determination of molecules or the bond behavior in samples. A new SERS method was proposed to get the spectrum of tea species to detect the sample purity involved in different types of planting and processing. The fingerprint SERS spectrum of seven kinds of tea samples was obtained. Data processing method of Principal Component Analysis (PCA) was used to separate tea species and several models for different tea samples. The combined method of fingerprinting-PCA was accurate and rapid for the evaluation of different tea species [112,113].

Pollen extracts from various plant species have different beneficial biological effects [114]. Seifert et al. [115,116] analyzed aqueous extracts from different kinds of pollen using SERS with AuNPs substrate. The SERS spectra of targets were specific, and the different species could be distinguished to classify the genus. The accurate identification of pollen was achieved by analyzing the intrinsic information in SERS data. Results showed that SERS had good potential to characterize and identify pollen species, and could improve the study of pollen physiology. Routine investigations of food composition and vitamin/nutrient contents are challenged by food matrix complexity and low analyte content in samples. Radu et al. [117] simultaneously detected two B-vitamins using the SERS fingerprint method, which could sensitively identify analyte molecules at a low cost.

### *3.5. Other Plant Components*

SERS spectra of DNA can offer an assessment of the genetic identity of different kinds of plants [118,119]. Knowledge of genetic resources with high diversity is a precondition for developing novel species. Muntean et al. measured the half bandwidths of SERS of genomic DNAs in vitro-grown tissues of apple leaf. Results showed that the SERS method could be applied to studying the dynamics of DNA approaching the surface of the metallic substrate, with good perspectives of analyzing interactions of DNA-ligand or changes of DNA structure under environmental stress conditions [120,121]. They extended the application of this SERS method to other plant leaves, such as chrysanthemum, common sundew, edelweiss and so on [122]. The SERS spectra of genomic DNAs from tomato plants was also collected and the structural changes of DNAs undergoing cryopreservation were discussed [123]. Based on these works, they put forward that interactions of plant DNA and ligand or the precision DNA structure when approaching metallic surface could be further studied using SERS.

SERS can be used to investigate samples with weak Raman signals, for example juices and pulp. Camerlingo et al. [124] studied apple juices and pulp to verify the existence of fructose and pectin, which were related to the quality evaluation of these products. A home-made substrate fabricated with a glass slide decorated with AuNPs was designed and applied to the SERS detection. The obtained SERS spectra with legible Raman features provided useful information for the characterization of products in food processing.

Peanuts are a main life-threatening food allergen [125,126]. Gezer et al. [127] applied a biodegradable SERS technique to detecting Ara h1, the main kind of allergen protein, with LOD of 0.14 mg/mL. By using anti-Ara h1 monoclonal antibodies to functionalize the sensor surface, high specificity was achieved.

Palanco et al. [128] used plasmonic structures of silver nanoaggregates or films to enhance the detection of the chemical components of an onion layer. Results showed a competitive adsorption of molecules of onion and reporter. Different spectra from different parts of the layer indicated the complicated molecule structure of the plant. Shen et al. [129] presented nondestructive imaging of the living leaf using micro-Raman spectroscopy by delivering the carbon-encapsulated SERS tags into the living leaf. In vivo SERS spectra were used to investigate the distribution of tags, which could avoid interfering from autofluorescence. The novel modality of imaging provided SERS attractive potential for noninvasive biochemical imaging of living plants.

Cepeda-Perez et al. [130] reported the distribution and interaction of quantum dots (QDs) in the microalgae extracellular matrix. Changes in the Raman spectra of *Haematococcus pluvialis* microalgae caused by the adsorption of QDs were found by applying nano-sensors with bare anisotropic gold structures for SERS effect. This research demonstrated early QDs accumulation in plant cells which would benefit understanding of the environmental influence [131].

## 4. Application on the Detection of Microorganism Original Biomolecules

### 4.1. Bacteria Original Biomolecules

Sensitive and accurate pathogen detection is a key measure for ensuring public health due to the rapidly increasing infectious disease rate globally. The correct identification of pathogens in clinical or food samples assures the proper selection of clinical treatments or food safety procedures [132]. SERS is well-suited for bacteria detection from molecular to cellular level due to its sensitivity, selectivity, and compatiblity with other techniques. A representative work was presented by Meng et al. [133] in which a new type of SERS chip, consisting a sophisticated sandwich graphene (G)-AgNP-silicon (Si) nanohybrids, has been developed. The chip system could achieve both molecular detection and cellular analysis in samples. A schematic illustration of the developed platform was showed in Figure 5. The chip could realize sensitive and accurate quantification of adenosine triphosphate (ATP), with LOD of about 1 pM, and can also simultaneously capture, discriminate, and inactivate the bacteria. The efficiency of bacteria capture was 54% at the bacteria concentration of $10^8$ CFU/mL, and 93% antibacterial rate could be reached 24 h after treating.

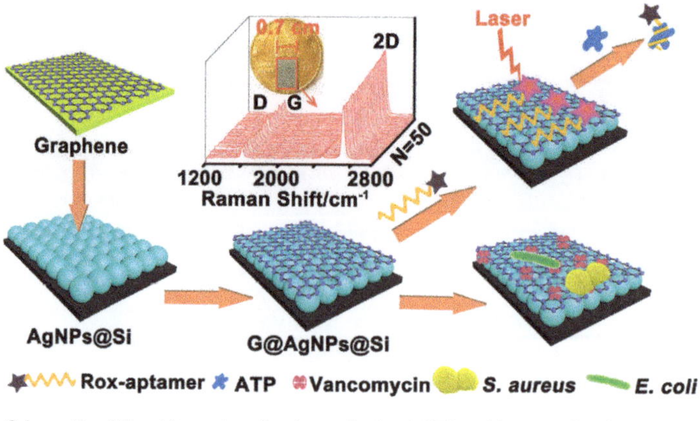


**Figure 5.** Schematic of the chip system for the analysis of ATP and bacteria [133]. Reproduced with permission from [133]. American Chemical Society, 2018.

For the application of SERS in the bacteria sensing, most of the studies focus on the cell detection, which is through the recognition of biomolecule on the cell membrane or inside the cell using antibodies, molecularly imprinted polymers or aptamers. The bacteria SERS signal can be significantly enhanced by these specifically recognized molecules, and the bacteria could be identified directly and visually from the SERS spectra [134]. There have been several reviews about the bacteria cell detection using SERS. Efrima et al. [135] reviewed studies using SERS for bacteria identification based on analyzing the spectra according to the nature of active centers and their distribution in the bacterium. Chauvet et al. [136] reviewed and proposed main strategies, such as methods to prepare the sample, from the bacterial culture conditions to the analysis of the spectra over the last 20 years. These reviews introduced the most recent reports about SERS detection of bacteria cells, which were comprehensive and detailed. So here the application of SERS on the detection of bacteria original biomolecules, rather than cells, was reviewed.

#### 4.1.1. Bacteria DNAs

Enteritidis caused by *Salmonella enterica* is a common foodborne disease increasingly rising globally [137]. Draz et al. [138] established an integrated assay for DNA detection of *Salmonella*. Using Au-nanoprobes, the LOD of the developed method (66 CFU/mL) was about 100-fold lower than the conventional PCR method.

Gracie et al. reported a novel detection method of three meningitis pathogens using lambda-Exonuclease digestion of double-stranded DNA and SERS detection. Two complementary DNA probes were simultaneously hybridized to a target sequence. The obtained double stranded DNA was digested using lambda-exonuclease and then detected by SERS. Meningitis pathogens were detected with LOD in the range of pico-molar [139].

Another simple and low-cost platform was designed to sensitively detect bacterial DNA by SERS using AuNPs modified as reporter probes, by using the separation and enrichment function of magnetic beads. A good linear relationship was gained of the DNA concentration range of 5 pM to 5 nM. The LOD for the detection of bacterial DNA was about 5 pM [140].

#### 4.1.2. Bacteria Proteins

Wang et al. [141] adopted a newly produced nanoyeast single-chain variable fragment to replace antibody, which is specific, cost-effective, and stable. By combining SERS with a microfluidic chip using nanoparticle clusters as labels, a universal platform for the sensitive and specific detection of pathogen antigens was established. LODs were 1 pg/mL for *Entamoeba histolytica* antigen EHI 115350 and 10 pg/mL for EHI 182030.

Trypsin shaving, as a targeted proteomic method, can be used to identify bacterial proteins exposed on cell-surface. For the redox-active proteins, obtained datasets were matched with SERS to identify the cofactors relevant with the cell-surface proteins. Further, this method could help solve problems concerning the existence of electron transport molecules in bacteria, especially microorganism that oxidize metals or metalloids [142].

#### 4.1.3. Other Bacteria Component Molecules

The bacterial outer membrane is composed of biochemical compounds that have specific information about bacterial strains, different stages of growth, responses to stimulation and so on [143]. Xu et al. [144] proposed that the molecule information of the bacterial outer membrane could be applied in rapidly detecting and identifying bacteria using the SERS method. Seven strains of the marine pathogen were used as models. Based on the SERS spectra, barcodes were generated for the detection of individual bacterial strains in blind samples. The developed sensing methods had broad applications in the areas of biomedical diagnostics, environmental monitoring, and security aspects. Just as proposed in the above report, SERS is a rapid and sensitive method with the potential to detect chemical changes on the surface of bacterial cell induced by the environmental changes. Stephe et al. [145] classified fourteen Arthrobacter strains with up to 97% accuracy by adopting PCA in combination with Linear Discriminant Analysis. Results showed that SERS could be used as a valuable tool to monitor and characterize phenotypic variations when faced with different environmental circumstances. A lab-on-a-chip (LOC)-SERS device was developed to differentiate six species of *mycobacteria*. The easy and reliable system was fabricated by the combination of a bead-beating module for cell disruption with the LOC-SERS device. Without analyte extraction or other treatments, the SERS spectra can quantify mycolic acid as the cell-wall component. By recording a data set using the LOC-SERS device, the type differentiation could also be achieved. At least 2100 SERS spectra could be obtained in 1 h [146]. Moreover, a reconfigurable assay was proposed to identify and monitor bacteria by direct detection of bacterial volatile organic compounds via SERS. Highly clinically relevant organisms were used to distinguish the species of bacteria with LOD of *Escherichia coli* at 10 CFU/mL in 12 h [147].

#### 4.1.4. Bacteria Metabolites

Determination of the chemical composition of biofilm matrices is crucial in different biological fields. The information of biofilm development and composition will help to select proper eliminating measures. Quorum sensing is significant in the survival of bacteria in biofilms and can be revealed by detecting related signaling metabolites. Multifunctional platforms for real-time tracing of metabolites secretion in biofilms were desired. Guo et al. [148] made a flexible and sticky sandwich note with two pieces of hexagonal boron nitride layers packaging gold nanostars, which can stick on natural biofilms for metabolites monitoring by SERS imaging sensitively. The sticky note can accurately quantify *Pseudomonas aeruginosa* after 1 h growth of biofilm by using its pyocyanin secretion as an internal standard for SERS signal self-calibration. This universal SERS sticky note can be used as a versatile tool in bacterial behavior research.

The metabolites of purine degradation excited at 785 nm are the major molecular species dominating the SERS spectra of bacteria. These molecules are produced by the bacterial starvation response in pure water washes following enrichment in culture mediums. The enzymes of bacterial supernatant that plays a main role in the known purine metabolism pathways were detected by SERS spectra to determine the bacterial specificity. These results showed that SERS could be a rapid diagnostic tool for metabolic profiling [149].

Zukovskaja et al. [150] developed a microfluidic device combined with SERS to analyze a *Pseudomonas aeruginosa* specific metabolite in aqueous solution with a LOD of 0.5 μM [151,152]. A simple and novel SERS platform is fabricated for in situ monitoring the nitric oxide (NO) release of a single bacterium. NO released under antibiotics and co-infected bacteria stress was investigated using this method [153]. *Staphylococcal* Enterotoxin B (SEB) was detected ultrasensitively using the SRES sandwich immunoassay. The antibody and SERS reporter molecule modified magnetic gold nanorod particles were used to capture SEB. A good linear relativity between the SEB concentration and SERS signal was found and the LOD was 768 mM [154].

For the SERS detection of different types of bacterial component molecules, more sophisticated nanomaterials, complicated microfluidic chips [133], and new types of recognition molecules such as nanoyeast single-chain variable fragment [141], were designed, fabricated, and used to prepare the SERS-active substrate.

### 4.2. *Virus Original Biomolecules*

Virus infection leads to severe epidemics of a large number of worldwide populations each year with high morbidity and mortality. Sensitive and simple detection of viruses is vital for control of viral spread at an early stage [155]. Virus detection methods are usually based on the immunoassay and PCR [156,157]. Considering the fact that they are rapid, portable and sensitive, the SERS-based immunoassay or PCR methods have potential application as a POC test in diagnosis.

### 4.2.1. Virus Proteins

Many kinds of virus surface proteins, such as G protein and antigen, were used to identify virus [158,159]. Lim et al. [160] tested the hypothesis that surface proteins and lipids of newly presented influenza viruses could enhance Raman peaks on AuNPs, which could then be distinguished from those of pseudotype with a noninfluenza virus component (Figure 6). This work provided a powerful label-free SERS platform to rapidly identify emerging influenza viruses.

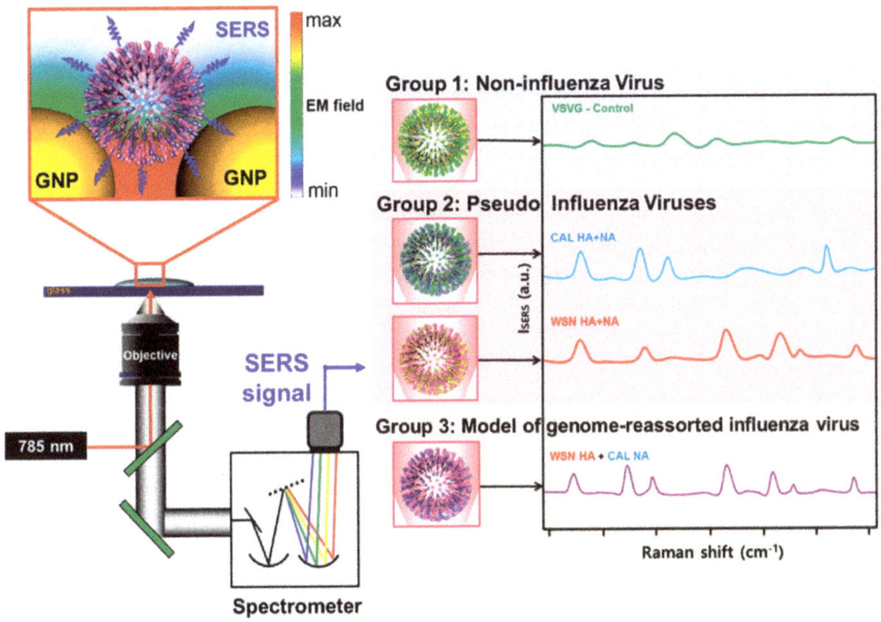


**Figure 6.** Schematic of the rapid detection of influenza viruses via SERS [160]. Reproduced with permission from [160]. American Chemical Society, 2015.

A sensitive immunoassay with SERS detection combined with microfluidic devices was developed using a novel Raman reporter molecule and substrate. Basic fuchsin (FC) was designed as Raman reporter, which can notably enhance SERS signal and connect the antibody and gold nanostructures. A good linear relativity between the intensity of SERS signal of FC band and concentration of Hepatitis B virus antigen was obtained with LOD of 0.01 IU/mL [161].

Sun et al. developed a magnetic immunosensor with SERS detection of intact and inactivated influenza virus H3N2 by fabricating a sandwich complex combining SERS tags, target viruses and supporting substrates. Using a portable Raman spectrometer, a good linear relationship was obtained with a LOD of $10^2$ TCID50/mL [162].

Porcine circovirus is a ubiquitous and crucial infectious virus in global pig farms [163,164]. Luo et al. reported an immunoassay combined with SERS detection for porcine circovirus type 2 (PCV2) using multi-branched mb-AuNPs combined with the PCV2 cap protein antibody as substrates and Raman reporters. A calibration curve plotting the intensity Raman signal at 1076 $cm^{-1}$ versus the concentrations of PCV2 was obtained from $8 \times 10^2$ to $8 \times 10^6$ copy/mL with the LOD of $8 \times 10^2$ copy/mL. This method is rapid, facile and sensitive compared to conventional PCR method [165]. They further used porous carbon films coated with AgNPs to determine PCV2, porcine parvovirus (PPV) and porcine pseudorabies virus (PRV). The LOD was improved as low as $1 \times 10^7$ copy/mL [166]. Enterovirus 71 (EV71), another health hazard, needs to be monitored with rapid POC detection. Reyes et al. developed a SERS method utilizing colloidal gold nanostars (AuNS) aggregation induced by protein for rapid detection of EV71 without substrates, Raman labels or sample preparing. AuNS was modified with scavenger receptor class B, member 2 (SCARB2) protein. In the absence of EV71, AuNS modified with scavenger receptor class B member 2 (SCARB2) protein aggregated produced four enhanced Raman peaks at 390, 510, 670, and 910 $cm^{-1}$. In the presence of EV71, as the virus bound to AuNS-SCARB2 preventing aggregation, the peak at 390 $cm^{-1}$ diminished in intensity with other three peaks disappeared, which could be potential indicators for the specific detection of EV71. EV71 could be detected in protein-rich medium within 15 min with this facile approach [167]. Sanchez-Purra et al. [168] explored the high sensitivity of SERS in a universal approach that could distinguish Zika from dengue nonstructural protein 1 (NS1) biomarkers. SERS-encoded gold nanostars were modified with the antibodies of both viruses for a dipstick immunoassay with 15-fold and 7-fold lower LOD for Zika NS1 and dengue NS1, respectively.

#### 4.2.2. Virus DNAs

With the distinguished spectral properties of metal carbonyls, Lin et al. [169] proposed a SERS ratiometric assay to detect cell-free circulating DNA (cfDNA) derived from the Epstein-Barr virus in human blood samples for nasopharyngeal cancer. Rhenium carbonyl (Re-CO) was used as a DNA probe, and osmium carbonyl (Os-CO) was used as an internal reference. The binding of Re-CO to cfDNA is accompanied with a performance of a stretching vibrations peak at 2113 $cm^{-1}$ overlapping with Os-CO (2025 $cm^{-1}$). This led to an increase in the ratio of I-2113/I-2025, quantitatively corresponding to the increase of cfDNA. The SERS method can be applied to detecting cfDNA in clinical blood samples because the ratio of I-2113/I-2025 lying in the region of 1780–2200 $cm^{-1}$ of the biomolecules.

An indirect capture method was established using colloidal AuNPs for SERS detection of DNA. The capture sequence obtained from the RNA genome of West Nile Virus. Colloidal gold was modified with a complementary capture oligonucleotide and a reporter oligonucleotide combined with methylene blue as Raman label. The LOD was in the submicromolar range [170].

To obtain sensitive, stable and reproducible gene detection of respiratory syncytial virus (RSV), a new SERS substrates were fabricated by electroless metallization of Ag and vapor phase deposition of Au on the nanostructured templates. Gene detection of RSV was achieved with a molecular probe with a fluorescent moiety and a linker to be attached on the substrate. To detect multiple targets, molecular probes were designed using a broad range of fluorophores. This reproducible dual-mode method (i.e., fluorescent and SERS) was self-confirmatory and can eliminate false positives [171].

For the SERS detection of virus biomolecules, the newly developed free-substrate method based on the AuNS aggregation induced by protein was used to achieve the rapid detection of EV71 [167]. Label-free SERS platform, microfluidic chip with new structure and new type of DNA probe were also applied to detect virus biomolecules.

### 4.3. *Other Microorganism Original Biomolecules*

Fungal infections cause high morbidity and mortality among hospitalized patients and immunocompromised individuals; fast and accurate diagnosis for fungal diseases is in demand [172]. SERS combined with PCA was used to detect and identify human fungal pathogens rapidly and reliably. Dina et al. [173] distinguished different clinical samples of fungal species using the chemometrics assisted SERS-based method. The overall analysis of the SERS spectra was carried out using appropriate chemometric tools-classical and fuzzy PCA combined with linear discriminant analysis to analysis the first principal components. Discrimination between several species of fungal pathogen strains showed that the established method could be applied as an alternative routine analysis tool in clinical diagnosis. Witkowska et al. [174] confirmed that the SERS method could effectively distinguish between certain fungal pathogens and offer taxonomic relation of fungi. Moreover, using the PCA analysis, statistical classification of fungi could be performed. Two principal components calculated were the most clinically significant, displaying 97% of the variability and discriminate between fungal species.

Zivanovic et al. [175] reported the molecular composition probing of Leishmania-infected macrophage cells by SERS. The data was used to assess the distribution of cholesterol and ergosterolin the amastigote period of the parasite and its vacuole ambient enviroment. Parasite original proteophosphoglycans, an important infection marker, were identified. Mycoplasma pneumoniae, as a respiratory related pathogen, can cause chronic bronchitis and pneumonia. The main surface protein P1 needed to form complexes with certain proteins to act in receptor binding or motility, and the variability in the related proteins can be used to distinguish the major genotypes. Strains with different genotype can be discriminated sensitively and specifically by using SERS on silver nanorod arrays. Krause et al. [176] applied the variable selection method to the identification of Raman bands vital in the classification of *M. pneumoniae* strains. The current methods for malaria diagnosis are time consuming, and are not suitable for early disease diagnosis. Garrett et al. [177] developed a reliable method based on a novel gold-coated SERS substrate and applied to the detection of malarial hemozoin pigment in the blood samples with 0.005% and 0.0005% infected cells.

## 5. Conclusions and Outlook

The recent explosive growth of the SERS studies brings these sensitive, rapid, accurate and reliable methods to more application fields. The merits of SERS expand its potential in the identification and quantification of various Raman-active biomolecules. Based on the development of new SERS strategies and combined with other techniques (e.g., immunoassay, PCR, and microfluidic chip), SERS shows increasingly broad application in detecting biomolecules from humans, animals, plants, and microorganisms, which can contribute to improving food safety, clinical diagnosis, and environmental monitoring. To be specific, SERS spectrum is able to provide abundant clues about the structure and/or quantity of interested molecules, including the hazard(s) or nutrient(s) present in foods, the pollutants in environmental samples, pathogens in clinical samples or biological processes occurring at the cellular or molecular level, etc.

This review article provided an overview of the recent progress and current shortages of SERS on biomolecule detection in order to indicate the application direction in the future. For example, there is a lack of and an urgent need for SERS application in the basic study of life sciences, such as animal or plant physiology that is crucial for getting more deep and basic knowledge on food security, environmental aspects and the ecological system. Therefore, the target biomolecules in life sciences were selected in this review to attract the attention of experts in both SERS and life science research and to expand the depth and fields of SERS application. There are still some technological problems that urgently require to be solved. Robust and reproducible SERS substrates as the critical parts of SERS methods are still under development, and the property needs to be improved and adapted to a wider range of analytes and biological samples. New types of nanomaterials that can improve the sensitivity, specificity and stability of SERS detection will benefit from the progress in nanotechnology and nanofabrication. More and more sophisticated nanostructures will be successfully fabricated and the problems of inconsistency and production cost of using SERS-active substrates will be gradually solved. The application of SERS in the detection of biology molecules was also seriously limited and challenged by the complex matrix interference of biology samples. The development of more effective sample preparation and purification methods that can be integrated with SERS detection would allow for a wider application range of SERS detection of biological molecules. To achieve the efficient extraction and purification of target analytes, the selection of specific biological or biomimetic recognition molecules, such as antibody, molecularly imprinted polymers, and aptamer with strong affinity and specificity, and their application on the modification of SERS substrates will be important technological factors. Furthermore, through the use of flexible substrates (e.g., paper and plastic films), sensitive, low cost and disposable commercial SERS platforms will be developed to adapt to POC applications in resource-limited settings. Finally, combined with the development of nanotechnology and biochemical methodology, SERS holds great promise in showing superior capabilities in biological fields in future.

**Author Contributions:** M.J. and H.Z. outlined and mainly wrote the manuscript; X.L., S.L. and L.Z. partially supervised the manuscript process. All authors contributed to the discussion, reviewed the manuscript, and accepted its final form.

**Funding:** This work was funded by National Key R&D Program of China (2016YFD0401101), the National Natural Science Foundation of China (31871874), the Natural Science Foundation of Shandong Province (ZR2016CM47), and the State Scholarship Fund by the China Scholarship Council (File No. 201708370074).

**Conflicts of Interest:** The authors declare no conflict of interest.

## References

1. Li, L.; Yang, H.J.; Liu, D.C.; He, H.B.; Wang, C.F.; Zhong, J.F.; Gao, Y.D.; Zeng, Y.J. Analysis of biofilms formation and associated genes detection in staphylococcus isolates from bovine mastitis. *Int. J. Appl. Res. Vet. Med.* **2012**, *10*, 62–68. [CrossRef]
2. Xie, S.B.; Zhou, J. Harnessing plant biodiversity for the discovery of novel anticancer drugs targeting microtubules. *Front. Plant Sci.* **2017**, *8*. [CrossRef] [PubMed]
3. Zhou, L.; Zhou, J.; Feng, Z.; Wang, F.Y.; Xie, S.S.; Bu, S.Z. Immunoassay for tumor markers in human serum based on si nanoparticles and sic@ag sers-active substrate. *Analyst* **2016**, *141*, 2534–2541. [CrossRef] [PubMed]
4. Zhang, Z.; Wang, Q.; Han, L.; Du, S.; Yu, H.; Zhang, H. Rapid and sensitive detection of salmonella typhimurium based on the photothermal effect of magnetic nanomaterials. *Sens. Actuators B Chem.* **2018**, *268*, 188–194. [CrossRef]
5. Rombout, J.; Yang, G.W.; Kiron, V. Adaptive immune responses at mucosal surfaces of teleost fish. *Fish Shellfish Immunol.* **2014**, *40*, 634–643. [CrossRef] [PubMed]
6. Sui, N.; Wang, Y.; Liu, S.S.; Yang, Z.; Wang, F.; Wan, S.B. Transcriptomic and physiological evidence for the relationship between unsaturated fatty acid and salt stress in peanut. *Front. Plant Sci.* **2018**, *9*. [CrossRef] [PubMed]
7. Hanif, S.; Liu, H.L.; Chen, M.; Muhammad, P.; Zhou, Y.; Cao, J.; Ahmed, S.A.; Xu, J.J.; Xia, X.H.; Chen, H.Y.; et al. Organic cyanide decorated sers active nanopipettes for quantitative detection of hemeproteins and fe3+ in single cells. *Anal. Chem.* **2017**, *89*, 2522–2530. [CrossRef] [PubMed]
8. Black, D.M.; Robles, G.; Lopez, P.; Bach, S.B.H.; Alvarez, M.; Whetten, R.L. Liquid chromatography separation and mass spectrometry detection of silver-lipoate ag29(la)12 nanoclusters: Evidence of isomerism in the solution phase. *Anal. Chem.* **2018**, *90*, 2010–2017. [CrossRef] [PubMed]
9. Sun, Z.; Lv, J.; Liu, X.; Tang, Z.; Wang, X.; Xu, Y.; Hammock, B.D. Development of a nanobody-avitag fusion protein and its application in a streptavidin-biotin-amplified enzyme-linked immunosorbent assay for ochratoxin a in cereal. *Anal. Chem.* **2018**. [CrossRef] [PubMed]
10. Deng, Y.Q.; Feng, Z.T.; Yuan, F.; Guo, J.R.; Suo, S.S.; Wang, B.S. Identification and functional analysis of the autofluorescent substance in limonium bicolor salt glands. *Plant Physiol. Biochem.* **2015**, *97*, 20–27. [CrossRef] [PubMed]
11. Yuan, F.; Chen, M.; Leng, B.Y.; Wang, B.S. An efficient autofluorescence method for screening limonium bicolor mutants for abnormal salt gland density and salt secretion. *S. Afr. J. Bot.* **2013**, *88*, 110–117. [CrossRef]
12. Guo, F.D.; Liu, C.L.; Xia, H.; Bi, Y.P.; Zhao, C.Z.; Zhao, S.Z.; Hou, L.; Li, F.G.; Wang, X.J. Induced expression of atlec1 and atlec2 differentially promotes somatic embryogenesis in transgenic tobacco plants. *PLoS ONE* **2013**, *8*. [CrossRef] [PubMed]
13. Li, X.Z.; Yang, T.Y.; Li, C.S.; Jin, L.L.; Lou, H.; Song, Y.T. Prenatal detection of thalassemia by cell-free fetal DNA (cffdna) in maternal plasma using surface enhanced raman spectroscopy combined with pcr. *Biomed. Opt. Express* **2018**, *9*, 3167–3176. [CrossRef] [PubMed]
14. Liu, R.M.; Xiong, Y.; Tang, W.Y.; Guo, Y.; Yan, X.H.; Si, M.Z. Near-infrared surface-enhanced raman spectroscopy (nir-sers) studies on oxyheamoglobin (oxyhb) of liver cancer based on pva-ag nanofilm. *J. Raman Spectrosc.* **2013**, *44*, 362–369. [CrossRef]

15. Casella, M.; Lucotti, A.; Tommasini, M.; Bedoni, M.; Forvi, E.; Gramatica, F.; Zerbi, G. Raman and sers recognition of beta-carotene and haemoglobin fingerprints in human whole blood. *Spectrochim. Acta Part A Mol. Biomol. Spectrosc.* **2011**, *79*, 915–919. [CrossRef] [PubMed]
16. Wang, Z.; Yang, H.Q.; Wang, M.H.; Petti, L.; Jiang, T.; Jia, Z.H.; Xie, S.S.; Zhou, J. Sers-based multiplex immunoassay of tumor markers using double sio2@ag immune probes and gold-film hemisphere array immune substrate. *Colloids Surf. A Physicochem. Eng. Asp.* **2018**, *546*, 48–58. [CrossRef]
17. Liu, N.N.; Chi, Y.N.; Dong, L.H.; Li, J.H.; Shan, G.Y.; Chen, Y.W.; Liu, Y.C. Label-free detection of bovine serum albumin protein based on sio2/au nanoshells as near-infrared surface-enhanced raman spectroscopy nanoprobe. *J. Nanosci. Nanotechnol.* **2016**, *16*, 7103–7109. [CrossRef]
18. Liang, A.H.; Li, C.N.; Wang, X.L.; Luo, Y.H.; Wen, G.Q.; Jiang, Z.L. Immunocontrolling graphene oxide catalytic nanogold reaction and its application to sers quantitative analysis. *ACS Omega* **2017**, *2*, 7349–7358. [CrossRef] [PubMed]
19. Jin, B.B.; He, J.H.; Li, J.; Zhang, Y. Lotus seedpod inspired sers substrates: A novel platform consisting of 3d sub-10 nm annular hot spots for ultrasensitive sers detection. *Adv. Opt. Mater.* **2018**, *6*. [CrossRef]
20. Yang, Y.J.; Jiang, X.Y.; Chao, J.; Song, C.Y.; Liu, B.; Zhu, D.; Sun, Y.Z.; Yang, B.Y.; Zhang, Q.W.; Chen, Y.; et al. Synthesis of magnetic core-branched au shell nanostructures and their application in cancer-related mirna detection via sers. *Sci. China Mater.* **2017**, *60*, 1129–1144. [CrossRef]
21. Karadan, P.; Aggarwal, S.; Anappara, A.A.; Narayana, C.; Barshilia, H.C. Tailored periodic si nanopillar based architectures as highly sensitive universal sers biosensing platform. *Sens. Actuators B Chem.* **2018**, *254*, 264–271. [CrossRef]
22. David, C.; Guillot, N.; Shen, H.; Toury, T.; de la Chapelle, M.L. Sers detection of biomolecules using lithographed nanoparticles towards a reproducible sers biosensor. *Nanotechnology* **2010**, *21*. [CrossRef] [PubMed]
23. Restaino, S.M.; White, I.M. Inkjet-printed paper surface enhanced raman spectroscopy (sers) sensors: Portable, low cost diagnostics for microrna. In Proceedings of the 2016 IEEE SENSORS, Orlando, FL, USA, 30 October–3 November 2016; pp. 1–3.
24. Li, L.; Liu, C.; Cao, X.W.; Wang, Y.; Dong, J.; Qian, W.P. Determination of carcinoembryonic antigen by surface-enhanced raman spectroscopy using gold nanobowl arrays. *Anal. Lett.* **2017**, *50*, 982–998. [CrossRef]
25. Yang, L.Y.; Fu, C.C.; Wang, H.L.; Xu, S.P.; Xu, W.Q. Aptamer-based surface-enhanced raman scattering (sers) sensor for thrombin based on supramolecular recognition, oriented assembly, and local field coupling. *Anal. Bioanal. Chem.* **2017**, *409*, 235–242. [CrossRef] [PubMed]
26. Foti, A.; D'Andrea, C.; Villari, V.; Micali, N.; Donato, M.G.; Fazio, B.; Marago, O.M.; Gillibert, R.; de la Chapelle, M.L.; Gucciardi, P.G. Optical aggregation of gold nanoparticles for sers detection of proteins and toxins in liquid environment: Towards ultrasensitive and selective detection. *Materials* **2018**, *11*. [CrossRef] [PubMed]
27. Zhou, J.H.; Ren, K.N.; Zhao, Y.H.; Dai, W.; Wu, H.K. Convenient formation of nanoparticle aggregates on microfluidic chips for highly sensitive sers detection of biomolecules. *Anal. Bioanal. Chem.* **2012**, *402*, 1601–1609. [CrossRef] [PubMed]
28. Kang, Y.P.; Si, M.Z.; Zhu, Y.Q.; Miao, L.; Xu, G. Surface-enhanced raman scattering (sers) spectra of hemoglobin of mouse and rabbit with self-assembled nano-silver film. *Spectrochim. Acta Part A Mol. Biomol. Spectrosc.* **2013**, *108*, 177–180. [CrossRef] [PubMed]
29. He, J.C.; Li, G.K.; Hu, Y.L. Aptamer recognition induced target-bridged strategy for proteins detection based on magnetic chitosan and silver/chitosan nanoparticles using surface-enhanced raman spectroscopy. *Anal. Chem.* **2015**, *87*, 11039–11047. [CrossRef] [PubMed]
30. Wu, Z.T.; Liu, Y.Z.; Zhou, X.D.; Shen, A.G.; Hu, J.M. A turn-off sers-based detection platform for ultrasensitive detection of thrombin based on, enzymatic assays. *Biosens. Bioelectron.* **2013**, *44*, 10–15. [CrossRef] [PubMed]
31. Meng, X.Q.; Dai, Y.Y.; Jing, L.D.; Bai, J.; Liu, S.Z.; Zheng, K.G.; Pan, J. Subcellular localization of proline-rich tyrosine kinase 2 during oocyte fertilization and early-embryo development in mice. *J. Reprod. Dev.* **2016**, *62*, 351–358. [CrossRef] [PubMed]
32. Liu, W.; Ji, S.X.; Fang, X.L.; Wang, Q.G.; Li, Z.; Yao, F.Y.; Hou, L.; Dai, S.J. Protein kinase ltrpk1 influences cold adaptation and microtubule stability in rice. *J. Plant Growth Regul.* **2013**, *32*, 483–490. [CrossRef]

33. Zhao, S.S.; Jiang, Y.X.; Zhao, Y.; Huang, S.J.; Yuan, M.; Zhao, Y.X.; Guo, Y. Casein kinase1-like protein2 regulates actin filament stability and stomatal closure via phosphorylation of actin depolymerizing factor. *Plant Cell* **2016**, *28*, 1422–1439. [CrossRef] [PubMed]
34. Cottat, M.; Yasukuni, R.; Homma, Y.; Lidgi-Guigui, N.; Varin-Blank, N.; de la Chapelle, M.L.; Le Roy, C. Phosphorylation impact on spleen tyrosine kinase conformation by surface enhanced raman spectroscopy. *Sci. Rep.* **2017**, *7*. [CrossRef] [PubMed]
35. He, S.; Kyaw, Y.M.E.; Tan, E.K.M.; Bekale, L.; Kang, M.W.C.; Kim, S.S.Y.; Tan, I.; Lam, K.P.; Kah, J.C.Y. Quantitative and label-free detection of protein kinase a activity based on surface-enhanced raman spectroscopy with gold nanostars. *Anal. Chem.* **2018**, *90*, 6071–6080. [CrossRef] [PubMed]
36. Zhu, Y.Y.; Xing, W.X.; Shan, S.J.; Zhang, S.Q.; Li, Y.Q.; Li, T.; An, L.; Yang, G.W. Characterization and immune response expression of the rig-i-like receptor mda5 in common carp cyprinus carpio. *J. Fish Biol.* **2016**, *88*, 2188–2202. [CrossRef] [PubMed]
37. Zhang, F.M.; Liu, D.Z.; Wang, L.; Li, T.; Chang, Q.; An, L.G.; Yang, G.W. Characterization of igm-binding protein: A pigr-like molecule expressed by intestinal epithelial cells in the common carp (cyprinus carpio l.). *Vet. Immunol. Immunopathol.* **2015**, *167*, 30–35. [CrossRef] [PubMed]
38. Li, T.; Li, H.; Peng, S.Q.; Zhang, F.M.; An, L.G.; Yang, G.W. Molecular characterization and expression pattern of x box-binding protein-1 (xbp1) in common carp (cyprinus carpio l.): Indications for a role of xbp1 in antibacterial and antiviral immunity. *Fish Shellfish Immunol.* **2017**, *67*, 667–674. [CrossRef] [PubMed]
39. Yang, H.T.; Zou, S.S.; Zhai, L.J.; Wang, Y.; Zhang, F.M.; An, L.G.; Yang, G.W. Pathogen invasion changes the intestinal microbiota composition and induces innate immune responses in the zebrafish intestine. *Fish Shellfish Immunol.* **2017**, *71*, 35–42. [CrossRef] [PubMed]
40. Penn, M.A.; Drake, D.M.; Driskell, J.D. Accelerated surface-enhanced raman spectroscopy (sers)-based immunoassay on a gold-plated membrane. *Anal. Chem.* **2013**, *85*, 8609–8617. [CrossRef] [PubMed]
41. Gu, X.F.; Yan, Y.R.; Jiang, G.Q.; Adkins, J.; Shi, J.; Jiang, G.M.; Tian, S. Using a silver-enhanced microarray sandwich structure to improve sers sensitivity for protein detection. *Anal. Bioanal. Chem.* **2014**, *406*, 1885–1894. [CrossRef] [PubMed]
42. Neng, J.; Harpster, M.H.; Zhang, H.; Mecham, J.O.; Wilson, W.C.; Johnson, P.A. A versatile sers-based immunoassay for immunoglobulin detection using antigen-coated gold nanoparticles and malachite green-conjugated protein a/g. *Biosens. Bioelectron.* **2010**, *26*, 1009–1015. [CrossRef] [PubMed]
43. Cui, L.L.; Yang, G.W.; Pan, J.; Zhang, C. Tumor necrosis factor alpha knockout increases fertility of mice. *Theriogenology* **2011**, *75*, 867–876. [CrossRef] [PubMed]
44. Shan, S.J.; Qi, C.C.; Zhu, Y.Y.; Li, H.; An, L.G.; Yang, G.W. Expression profile of carp ifn correlate with the up-regulation of interferon regulatory factor-1 (irf-1) in vivo and in vitro: The pivotal molecules in antiviral defense. *Fish Shellfish Immunol.* **2016**, *52*, 94–102. [CrossRef] [PubMed]
45. Kaminska, A.; Winkler, K.; Kowalska, A.; Witkowska, E.; Szymborski, T.; Janeczek, A.; Waluk, J. Sers-based immunoassay in a microfluidic system for the multiplexed recognition of interleukins from blood plasma: Towards picogram detection. *Sci. Rep.* **2017**, *7*. [CrossRef] [PubMed]
46. Kaminska, A.; Sprynskyy, M.; Winkler, K.; Szymborski, T. Ultrasensitive sers immunoassay based on diatom biosilica for detection of interleukins in blood plasma. *Anal. Bioanal. Chem.* **2017**, *409*, 6337–6347. [CrossRef] [PubMed]
47. Wang, Y.; Tang, L.J.; Jiang, J.H. Surface-enhanced raman spectroscopy-based, homogeneous, multiplexed immunoassay with antibody-fragments-decorated gold nanoparticles. *Anal. Chem.* **2013**, *85*, 9213–9220. [CrossRef] [PubMed]
48. Liu, T.T.; Liu, S.; Ma, L.; Li, F.L.; Zheng, Z.D.; Chai, R.F.; Hou, Y.H.; Xie, Y.B.; Li, G.R. Oogenesis, vitellogenin-mediated ovarian degeneration and immune response in the annual fish nothobranchius guentheri. *Fish Shellfish Immunol.* **2017**, *66*, 86–92. [CrossRef] [PubMed]
49. Wen, G.Q.; Liang, X.J.; Liu, Q.Y.; Liang, A.H.; Jiang, Z.L. A novel nanocatalytic sers detection of trace human chorionic gonadotropin using labeled-free vitoria blue 4r as molecular probe. *Biosens. Bioelectron.* **2016**, *85*, 450–456. [CrossRef] [PubMed]
50. Guo, T.; Zhang, L.; Cheng, D.; Liu, T.; An, L.G.; Li, W.P.; Zhang, C. Low-density lipoprotein receptor affects the fertility of female mice. *Reprod. Fertil. Dev.* **2015**, *27*, 1222–1232. [CrossRef] [PubMed]
51. Xiao, L.F.; Parchur, A.K.; Gilbertson, T.A.; Zhou, A.H. Sers-fluorescence bimodal nanoprobes for in vitro imaging of the fatty acid responsive receptor gpr120. *Anal. Methods* **2018**, *10*, 22–29. [CrossRef] [PubMed]

52. Chang, H.J.; Kang, H.M.; Ko, E.; Jun, B.H.; Lee, H.Y.; Lee, Y.S.; Jeang, D.H. Psa detection with femtomolar sensitivity and a broad dynamic range using sers nanoprobes and an area-scanning method. *ACS Sens.* **2016**, *1*, 645–649. [CrossRef]
53. Lin, Y.; Xu, G.H.; Wei, F.D.; Zhang, A.X.; Yang, J.; Hu, Q. Detection of cea in human serum using surface-enhanced raman spectroscopy coupled with antibody-modified au and gamma-fe2o3@au nanoparticles. *J. Pharm. Biomed. Anal.* **2016**, *121*, 135–140. [CrossRef] [PubMed]
54. Guo, M.D.; Dong, J.; Xie, W.; Tao, L.; Lu, W.B.; Wang, Y.; Qian, W.P. Sers tags-based novel monodispersed hollow gold nanospheres for highly sensitive immunoassay of cea. *J. Mater. Sci.* **2015**, *50*, 3329–3336. [CrossRef]
55. Gao, R.K.; Cheng, Z.Y.; Demello, A.J.; Choo, J. Wash-free magnetic immunoassay of the psa cancer marker using sers and droplet microfluidics. *Lab Chip* **2016**, *16*, 1022–1029. [CrossRef] [PubMed]
56. Guo, W.J.; Hu, Y.H.; Wei, H. Enzymatically activated reduction-caged sers reporters for versatile bioassays. *Analyst* **2017**, *142*, 2322–2326. [CrossRef] [PubMed]
57. Song, G.F.; Zhou, H.; Gu, J.J.; Liu, Q.L.; Zhang, W.; Su, H.L.; Su, Y.S.; Yao, Q.H.; Zhang, D. Tumor marker detection using surface enhanced raman spectroscopy on 3d au butterfly wings. *J. Mater. Chem. B* **2017**, *5*, 1594–1600. [CrossRef]
58. Karaballi, R.A.; Merchant, S.; Power, S.R.; Brosseau, C.L. Electrochemical surface-enhanced raman spectroscopy (ec-sers) study of the interaction between protein aggregates and biomimetic membranes. *Phys. Chem. Chem. Phys.* **2018**, *20*, 4513–4526. [CrossRef] [PubMed]
59. Sinha, S.S.; Jones, S.; Pramanik, A.; Ray, P.C. Nanoarchitecture based sers for biomolecular fingerprinting and label-free disease markers diagnosis. *Acc. Chem. Res.* **2016**, *49*, 2725–2735. [CrossRef] [PubMed]
60. Wang, Y.; Zhao, P.; Mao, L.; Hou, Y.; Li, D. Determination of brain injury biomarkers by surface-enhanced raman scattering using hollow gold nanospheres. *RSC Adv.* **2018**, *8*, 3143–3150. [CrossRef]
61. Li, D.; Yang, M.; Li, H.; Mao, L.; Wang, Y.; Sun, B. Sers based protocol flow glass-hemostix for detection of neuron-specific enolase in blood plasma. *New J. Chem.* **2018**. [CrossRef]
62. Gao, X.; Zheng, P.; Kasani, S.; Wu, S.; Yang, F.; Lewis, S.; Nayeem, S.; Engler-Chiurazzi, E.B.; Wigginton, J.G.; Simpkins, J.W.; et al. Paper-based surface-enhanced raman scattering lateral flow strip for detection of neuron-specific enolase in blood plasma. *Anal. Chem.* **2017**, *89*, 10104–10110. [CrossRef] [PubMed]
63. Wang, X.J.; Ju, Z.H.; Huang, J.M.; Hou, M.H.; Zhou, L.; Qi, C.; Zhang, Y.; Gao, Q.; Pan, Q.; Li, G.R.; et al. The relationship between the variants of the bovine mbl2 gene and milk production traits, mastitis, serum mbl-c levels and complement activity. *Vet. Immunol. Immunopathol.* **2012**, *148*, 311–319. [CrossRef] [PubMed]
64. Li, H.; Zhang, F.M.; Guo, H.Y.; Zhu, Y.Y.; Yuan, J.D.; Yang, G.W.; An, L.G. Molecular characterization of hepcidin gene in common carp (cyprinus carpio l.) and its expression pattern responding to bacterial challenge. *Fish Shellfish Immunol.* **2013**, *35*, 1030–1038. [CrossRef] [PubMed]
65. Shan, S.J.; Liu, D.Z.; Wang, L.; Zhu, Y.Y.; Zhang, F.M.; Li, T.; An, L.G.; Yang, G.W. Identification and expression analysis of irak1 gene in common carp cyprinus carpio l.: Indications for a role of antibacterial and antiviral immunity. *J. Fish Boil.* **2015**, *87*, 241–255. [CrossRef] [PubMed]
66. Girel, K.; Yantcevich, E.; Arzumanyan, G.; Doroshkevich, N.; Bandarenka, H. Detection of DNA molecules by sers spectroscopy with silvered porous silicon as an active substrate. *Phys. Status Solidi A Appl. Mater. Sci.* **2016**, *213*, 2911–2915. [CrossRef]
67. Han, B.; Zhang, Y.L.; Zhu, L.; Chen, X.H.; Ma, Z.C.; Zhang, X.L.; Wang, J.N.; Wang, W.; Liu, Y.Q.; Chen, Q.D.; et al. Direct laser scribing of agnps@rgo biochip as a reusable sers sensor for DNA detection. *Sens. Actuators B Chem.* **2018**, *270*, 500–507. [CrossRef]
68. Qian, Y.; Fan, T.T.; Yao, Y.; Shi, X.; Liao, X.J.; Zhou, F.Y.; Gao, F.L. Label-free and raman dyes-free surface-enhanced raman spectroscopy for detection of DNA. *Sens. Actuators B Chem.* **2018**, *254*, 483–489. [CrossRef]
69. Yu, J.; Jeon, J.; Choi, N.; Lee, J.O.; Kim, Y.P.; Choo, J. Sers-based genetic assay for amplification-free detection of prostate cancer specific pca3 mimic DNA. *Sens. Actuators B Chem.* **2017**, *251*, 302–309. [CrossRef]
70. Wang, R.; Kim, K.; Choi, N.; Wang, X.; Lee, J.; Jeon, J.H.; Rhie, G.-E.; Choo, J. Highly sensitive detection of high-risk bacterial pathogens using sers-based lateral flow assay strips. *Sens. Actuators B Chem.* **2018**, *270*, 72–79. [CrossRef]

71. Ye, S.J.; Wu, Y.Y.; Zhang, W.; Li, N.; Tang, B. A sensitive sers assay for detecting proteins and nucleic acids using a triple-helix molecular switch for cascade signal amplification. *Chem. Commun.* **2014**, *50*, 9409–9412. [CrossRef] [PubMed]
72. Kammer, E.; Olschewski, K.; Bocklitz, T.; Rosch, P.; Weber, K.; Cialla, D.; Popp, J. A new calibration concept for a reproducible quantitative detection based on sers measurements in a microfluidic device demonstrated on the model analyte adenine. *Phys. Chem. Chem. Phys.* **2014**, *16*, 9056–9063. [CrossRef] [PubMed]
73. Shi, M.L.; Zheng, J.; Tan, Y.J.; Tan, G.X.; Li, J.S.; Li, Y.H.; Li, X.; Zhou, Z.G.; Yang, R.H. Ultrasensitive detection of single nucleotide polymorphism in human mitochondrial DNA utilizing ion-mediated cascade surface-enhanced raman spectroscopy amplification. *Anal. Chem.* **2015**, *87*, 2734–2740. [CrossRef] [PubMed]
74. Cao, X.W.; Bao, M.; Shan, Y.B.; Li, W.; Shi, H.C. Rapid detection and identification of mirnas by surface-enhanced raman spectroscopy using hollow au nanoflowers substrates. *J. Nanomater.* **2017**. [CrossRef]
75. Driskell, J.D.; Tripp, R.A. Label-free sers detection of microrna based on affinity for an unmodified silver nanorod array substrate. *Chem. Commun.* **2010**, *46*, 3298–3300. [CrossRef] [PubMed]
76. Zheng, J.; Bai, J.H.; Zhou, Q.F.; Li, J.S.; Li, Y.H.; Yang, J.F.; Yang, R.H. DNA-templated in situ growth of agnps on swnts: A new approach for highly sensitive sers assay of microrna. *Chem. Commun.* **2015**, *51*, 6552–6555. [CrossRef] [PubMed]
77. Ye, S.; Wu, Y.; Zhai, X.; Tang, B. Asymmetric signal amplification for simultaneous sers detection of multiple cancer markers with significantly different levels. *Anal. Chem.* **2015**, *87*, 8242–8249. [CrossRef] [PubMed]
78. Zhou, W.; Tian, Y.F.; Yin, B.C.; Ye, B.C. Simultaneous surface-enhanced raman spectroscopy detection of multiplexed microrna biomarkers. *Anal. Chem.* **2017**, *89*, 6121–6129. [CrossRef] [PubMed]
79. Shin, K.; Choi, J.; Kim, Y.; Lee, Y.; Kim, J.; Lee, S.; Chung, H. Feasibility study for combination of field-flow fractionation (fff)-based separation of size-coded particle probes with amplified surface enhanced raman scattering (sers) tagging for simultaneous detection of multiple mirnas. *J. Chromatogr. A* **2018**, *1556*, 97–102. [CrossRef] [PubMed]
80. Liu, Y.J.; Xu, C.X.; Lu, J.F.; Zhu, Z.; Zhu, Q.X.; Manohari, A.G.; Shi, Z.L. Template-free synthesis of porous zno/ag microspheres as recyclable and ultra-sensitive sers substrates. *Appl. Surf. Sci.* **2018**, *427*, 830–836. [CrossRef]
81. Sooraj, K.P.; Ranjan, M.; Rao, R.; Mukherjee, S. Sers based detection of glucose with lower concentration than blood glucose level using plasmonic nanoparticle arrays. *Appl. Surf. Sci.* **2018**, *447*, 576–581. [CrossRef]
82. Qi, G.H.; Jia, K.Q.; Fu, C.C.; Xu, S.P.; Xu, W.Q. A highly sensitive sers sensor for quantitative analysis of glucose based on the chemical etching of silver nanoparticles. *J. Opt.* **2015**, *17*. [CrossRef]
83. Fu, C.; Jin, S.; Oh, J.; Xu, S.P.; Jung, Y.M. Facile detection of glucose in human serum employing silver-ion-guided surface-enhanced raman spectroscopy signal amplification. *Analyst* **2017**, *142*, 2887–2891. [CrossRef] [PubMed]
84. Tang, L.J.; Li, S.; Han, F.; Liu, L.Q.; Xu, L.G.; Ma, W.; Kuang, H.; Li, A.K.; Wang, L.B.; Xu, C.L. Sers-active au@ag nanorod dimers for ultrasensitive dopamine detection. *Biosens. Bioelectron.* **2015**, *71*, 7–12. [CrossRef] [PubMed]
85. Dumont, E.; De Bleye, C.; Cailletaud, J.; Sacre, P.Y.; Van Lerberghe, P.B.; Rogister, B.; Rance, G.A.; Aylott, J.W.; Hubert, P.; Ziemons, E. Development of a sers strategy to overcome the nanoparticle stabilisation effect in serum-containing samples: Application to the quantification of dopamine in the culture medium of pc-12 cells. *Talanta* **2018**, *186*, 8–16. [CrossRef] [PubMed]
86. Cao, X.M.; Qin, M.; Li, P.; Zhou, B.B.; Tang, X.H.; Ge, M.H.; Yang, L.B.; Liu, J.H. Probing catecholamine neurotransmitters based on iron-coordination surface-enhanced resonance raman spectroscopy label. *Sens. Actuators B Chem.* **2018**, *268*, 350–358. [CrossRef]
87. Moody, A.S.; Sharma, B. Multi-metal, multi-wavelength surface-enhanced raman spectroscopy detection of neurotransmitters. *ACS Chem. Neurosci.* **2018**, *9*, 1380–1387. [CrossRef] [PubMed]
88. Aguilar-Hernandez, I.; Afseth, N.K.; Lopez-Luke, T.; Contreras-Torres, F.F.; Wold, J.P.; Ornelas-Soto, N. Surface enhanced raman spectroscopy of phenolic antioxidants: A systematic evaluation of ferulic acid, *p*-coumaric acid, caffeic acid and sinapic acid. *Vib. Spectrosc.* **2017**, *89*, 113–122. [CrossRef]
89. Li, Y.; Driver, M.; Decker, E.; He, L.L. Lipid and lipid oxidation analysis using surface enhanced raman spectroscopy (sers) coupled with silver dendrites. *Food Res. Int.* **2014**, *58*, 1–6. [CrossRef]

90. Tang, G.Y.; Xu, P.L.; Ma, W.H.; Wang, F.; Liu, Z.J.; Wan, S.B.; Shan, L. Seed-specific expression of atlec1 increased oil content and altered fatty acid composition in seeds of peanut (arachis hypogaea l.). *Front. Plant Sci.* **2018**, *9*. [CrossRef] [PubMed]
91. Sui, N.; Li, M.; Li, K.; Song, J.; Wang, B.S. Increase in unsaturated fatty acids in membrane lipids of suaeda salsa l. Enhances protection of photosystem ii under high salinity. *Photosynthetica* **2010**, *48*, 623–629. [CrossRef]
92. Sun, Y.L.; Li, F.; Su, N.; Sun, X.L.; Zhao, S.J.; Meng, Q.W. The increase in unsaturation of fatty acids of phosphatidylglycerol in thylakoid membrane enhanced salt tolerance in tomato. *Photosynthetica* **2010**, *48*, 400–408. [CrossRef]
93. Sui, N.; Han, G.L. Salt-induced photoinhibition of psii is alleviated in halophyte thellungiella halophila by increases of unsaturated fatty acids in membrane lipids. *Acta Physiol. Plant.* **2014**, *36*, 983–992. [CrossRef]
94. Radu, A.I.; Ryabchykov, O.; Bocklitz, T.W.; Huebner, U.; Weber, K.; Cialla-May, D.; Popp, J. Toward food analytics: Fast estimation of lycopene and beta-carotene content in tomatoes based on surface enhanced raman spectroscopy (sers). *Analyst* **2016**, *141*, 4447–4455. [CrossRef] [PubMed]
95. Hsueh, H.Y.; Chen, H.Y.; Ling, Y.C.; Huang, W.S.; Hung, Y.C.; Gwo, S.; Ho, R.M. A polymer-based sers-active substrate with gyroid-structured gold multibranches. *J. Mater. Chem. C* **2014**, *2*, 4667–4675. [CrossRef]
96. Meng, X.; Yang, D.Y.; Li, X.D.; Zhao, S.Y.; Sui, N.; Meng, Q.W. Physiological changes in fruit ripening caused by overexpression of tomato slan2, an r2r3-myb factor. *Plant Physiol. Biochem.* **2015**, *89*, 24–30. [CrossRef] [PubMed]
97. Chan, T.Y.; Liu, T.Y.; Wang, K.S.; Tsai, K.T.; Chen, Z.X.; Chang, Y.C.; Tseng, Y.Q.; Wang, C.H.; Wang, J.K.; Wang, Y.L. Sers detection of biomolecules by highly sensitive and reproducible raman-enhancing nanoparticle array. *Nanoscale Res. Lett.* **2017**, *12*. [CrossRef] [PubMed]
98. Liu, S.S.; Wang, W.Q.; Li, M.; Wan, S.B.; Sui, N. Antioxidants and unsaturated fatty acids are involved in salt tolerance in peanut. *Acta Physiol. Plant.* **2017**, *39*. [CrossRef]
99. Wrona, M.; Salafranca, J.; Rocchia, M.; Nerin, C. Application of sers to the determination of butylated hydroxyanisole in edible and essential oils. *Spectroscopy* **2015**, *30*, 40–45.
100. Huang, C.C.; Chen, W.L. A sers method with attomolar sensitivity: A case study with the flavonoid catechin. *Microchim. Acta* **2018**, *185*. [CrossRef] [PubMed]
101. Liu, F.; Yang, Y.J.; Gao, J.W.; Ma, C.L.; Bi, Y.P. A comparative transcriptome analysis of a wild purple potato and its red mutant provides insight into the mechanism of anthocyanin transformation. *PLoS ONE* **2018**, *13*. [CrossRef] [PubMed]
102. Liu, J.; Zhang, F.; Zhou, J.J.; Chen, F.; Wang, B.S.; Xie, X.Z. Phytochrome b control of total leaf area and stomatal density affects drought tolerance in rice. *Plant Mol.Biol.* **2012**, *78*, 289–300. [CrossRef] [PubMed]
103. He, Y.A.; Li, Y.P.; Cui, L.X.; Xie, L.X.; Zheng, C.K.; Zhou, G.H.; Zhou, J.J.; Xie, X.Z. Phytochrome b negatively affects cold tolerance by regulating osdreb1 gene expression through phytochrome interacting factor-like protein ospil16 in rice. *Front. Plant Sci.* **2016**, *7*. [CrossRef] [PubMed]
104. Zhao, S.Z.; Sun, H.Z.; Chen, M.; Wang, B.S. Light-regulated betacyanin accumulation in euhalophyte suaeda salsa calli. *Plant Cell Tissue Organ Cult.* **2010**, *102*, 99–107. [CrossRef]
105. Zhao, S.Z.; Sun, H.Z.; Gao, Y.; Sui, N.; Wang, B.S. Growth regulator-induced betacyanin accumulation and dopa-4,5-dioxygenase (doda) gene expression in euhalophyte suaeda salsa calli. *In Vitro Cell. Dev. Biol. Plant* **2011**, *47*, 391–398. [CrossRef]
106. Zaffino, C.; Russo, B.; Bruni, S. Surface-enhanced raman scattering (sers) study of anthocyanidins. *Spectrochim. Acta Part A Mol. Biomol. Spectrosc.* **2015**, *149*, 41–47. [CrossRef] [PubMed]
107. Zaffino, C.; Bruni, S.; Russo, B.; Pilu, R.; Lago, C.; Colonna, G.M. Identification of anthocyanins in plant sources and textiles by surface-enhanced raman spectroscopy (sers). *J. Raman Spectrosc.* **2016**, *47*, 269–276. [CrossRef]
108. De Luca, E.; Redaelli, M.; Zaffino, C.; Bruni, S. A sers and hplc study of traditional dyes from native Chinese plants. *Vib. Spectrosc.* **2018**, *95*, 62–67. [CrossRef]
109. Gu, X.L.; Jin, Y.; Dong, F.; Cai, Y.Q.; You, Z.G.; You, J.H.; Zhang, L.Y.; Du, S.H. Toward rapid analysis, forecast and discovery of bioactive compounds from herbs by jointly using thin layer chromatography and ratiometric surface-enhanced raman spectroscopy technique. *J. Pharm. Biomed. Anal.* **2018**, *153*, 9–15. [CrossRef] [PubMed]

110. Zhang, W.; Wang, Y.P.; Bai, X.Y.; Zhao, B. Based on sers conformational studies of ginsenoside rb1 and its metabolites before and after combined with human serum albumin. *Spectrochim. Acta Part A Mol. Biomol. Spectrosc.* **2015**, *137*, 116–121. [CrossRef] [PubMed]
111. Zheng, J.K.; Fang, X.; Cao, Y.; Xiao, H.; He, L.L. Monitoring the chemical production of citrus-derived bioactive 5-demethylnobiletin using surface-enhanced raman spectroscopy. *J. Agric. Food Chem.* **2013**, *61*, 8079–8083. [CrossRef] [PubMed]
112. Buyukgoz, G.G.; Soforoglu, M.; Akgul, N.B.; Boyaci, I.H. Spectroscopic fingerprint of tea varieties by surface enhanced raman spectroscopy. *J. Food Sci. Technol. Mysore* **2016**, *53*, 1709–1716. [CrossRef] [PubMed]
113. Zhao, X.; Leavitt, S.D.; Zhao, Z.T.; Zhang, L.L.; Arup, U.; Grube, M.; Perez-Ortega, S.; Printzen, C.; Sliwa, L.; Kraichak, E.; et al. Towards a revised generic classification of lecanoroid lichens (lecanoraceae, ascomycota) based on molecular, morphological and chemical evidence. *Fungal Divers.* **2016**, *78*, 293–304. [CrossRef]
114. Huang, J.; Li, Z.Y.; Biener, G.; Xiong, E.H.; Malik, S.; Eaton, N.; Zhao, C.Z.; Raicu, V.; Kong, H.Z.; Zhao, D.Z. Carbonic anhydrases function in anther cell differentiation downstream of the receptor-like kinase ems1. *Plant Cell* **2017**, *29*, 1335–1356. [CrossRef] [PubMed]
115. Gong, W.P.; Han, R.; Li, H.S.; Song, J.M.; Yan, H.F.; Li, G.Y.; Liu, A.F.; Cao, X.Y.; Guo, J.; Zhai, S.N.; et al. Agronomic traits and molecular marker identification of wheat-aegilops caudata addition lines. *Front. Plant Sci.* **2017**, *8*. [CrossRef] [PubMed]
116. Seifert, S.; Merk, V.; Kneipp, J. Identification of aqueous pollen extracts using surface enhanced raman scattering (sers) and pattern recognition methods. *J. Biophotonics* **2016**, *9*, 181–189. [CrossRef] [PubMed]
117. Radu, A.I.; Kuellmer, M.; Giese, B.; Huebner, U.; Weber, K.; Cialla-May, D.; Popp, J. Surface-enhanced raman spectroscopy (sers) in food analytics: Detection of vitamins b-2 and b-12 in cereals. *Talanta* **2016**, *160*, 289–297. [CrossRef] [PubMed]
118. Zhao, C.Z.; Qiu, J.J.; Agarwal, G.; Wang, J.S.; Ren, X.Z.; Xia, H.; Guo, B.Z.; Ma, C.L.; Wan, S.B.; Bertioli, D.J.; et al. Genome-wide discovery of microsatellite markers from diploid progenitor species, arachis duranensis and a. Ipaensis, and their application in cultivated peanut (a. Hypogaea). *Front. Plant Sci.* **2017**, *8*. [CrossRef] [PubMed]
119. Wang, J.S.; Zhang, Q.; Cui, F.; Hou, L.; Zhao, S.Z.; Xia, H.; Qiu, J.J.; Li, T.T.; Zhang, Y.; Wang, X.J.; et al. Genome-wide analysis of gene expression provides new insights into cold responses in thellungiella salsuginea. *Front. Plant Sci.* **2017**, *8*. [CrossRef] [PubMed]
120. Yang, Z.; Wang, Y.; Wei, X.C.; Zhao, X.; Wang, B.S.; Sui, N. Transcription profiles of genes related to hormonal regulations under salt stress in sweet sorghum. *Plant Mol. Biol. Rep.* **2017**, *35*, 586–599. [CrossRef]
121. Muntean, C.M.; Leopold, N.; Halmagyi, A.; Valimareanu, S. Surface-enhanced raman spectroscopy of DNA from leaves of in vitro grown apple plants. *J. Raman Spectrosc.* **2011**, *42*, 844–850. [CrossRef]
122. Muntean, C.M.; Bratu, I.; Leopold, N.; Purcaru, M.A.P. Subpicosecond dynamics in DNA from leaves of in vitro-grown apple plants: A sers study. *Spectrosc. -Biomed. Appl.* **2011**, *26*, 59–68. [CrossRef]
123. Muntean, C.M.; Leopold, N.; Tripon, C.; Coste, A.; Halmagyi, A. Surface-enhanced raman spectroscopy of genomic DNA from in vitro grown tomato (lycopersicon esculentum mill.) cultivars before and after plant cryopreservation. *Spectrochim. Acta Part A Mol. Biomol. Spectrosc.* **2015**, *144*, 107–114. [CrossRef] [PubMed]
124. Camerlingo, C.; Portaccio, M.; Tate, R.; Lepore, M.; Delfino, I. Fructose and pectin detection in fruit-based food products by surface-enhanced raman spectroscopy. *Sensors* **2017**, *17*. [CrossRef] [PubMed]
125. Tang, G.Y.; Shao, F.X.; Xu, P.L.; Shan, L.; Liu, Z.J. Overexpression of a peanut nac gene, ahnac4, confers enhanced drought tolerance in tobacco. *Russ. J. Plant Physiol.* **2017**, *64*, 525–535. [CrossRef]
126. An, J.; Hou, L.; Li, C.; Wang, C.X.; Xia, H.; Zhao, C.Z.; Li, C.S.; Zheng, Y.X.; Zhao, Y.X.; Wang, X. Cloning and expression analysis of four della genes in peanut. *Russ. J. Plant Physiol.* **2015**, *62*, 116–126. [CrossRef]
127. Gezer, R.G.; Liu, G.L.; Kokini, J.L. Development of a biodegradable. Sensor platform from gold coated zein nanophotonic films to detect peanut allergen, ara h1, using surface enhanced raman spectroscopy. *Talanta* **2016**, *150*, 224–232. [CrossRef] [PubMed]
128. Palanco, M.E.; Mogensen, K.B.; Kneipp, K. Raman spectroscopic probing of plant material using sers. *J. Raman Spectrosc.* **2016**, *47*, 156–161. [CrossRef]
129. Shen, A.G.; Guo, J.Z.; Xie, W.; Sun, M.X.; Richards, R.; Hu, J.M. Surface-enhanced raman spectroscopy in living plant using triplex Au-Ag-C core-shell nanoparticles. *J. Raman Spectrosc.* **2011**, *42*, 879–884. [CrossRef]

130. Cepeda-Perez, E.; Aguilar-Hernandez, I.; Lopez-Luke, T.; Piazza, V.; Carriles, R.; Ornelas-Soto, N.; de la Rosa, E. Interaction of tga@cdte quantum dots with an extracellular matrix of haematococcus pluvialis microalgae detected using surface-enhanced raman spectroscopy (sers). *Appl. Spectrosc.* **2016**, *70*, 1561–1572. [CrossRef] [PubMed]
131. Casella, S.; Huang, F.; Mason, D.; Zhao, G.Y.; Johnson, G.N.; Mullineaux, C.W.; Liu, L.N. Dissecting the native architecture and dynamics of cyanobacterial photosynthetic machinery. *Mol. Plant.* **2017**, *10*, 1434–1448. [CrossRef] [PubMed]
132. Zhu, Y.Y.; Qi, C.C.; Shan, S.J.; Zhang, F.M.; Li, H.; An, L.G.; Yang, G.W. Characterization of common carp (cyprinus carpio l.) interferon regulatory factor 5 (irf5) and its expression in response to viral and bacterial challenges. *BMC Vet. Res.* **2016**, *12*. [CrossRef] [PubMed]
133. Meng, X.Y.; Wang, H.Y.; Chen, N.; Ding, P.; Shi, H.Y.; Zhai, X.; Su, Y.Y.; He, Y. A graphene-silver nanoparticle-silicon sandwich sers chip for quantitative detection of molecules and capture, discrimination, and inactivation of bacteria. *Anal. Chem.* **2018**, *90*, 5646–5653. [CrossRef] [PubMed]
134. Gao, W.C.; Li, B.; Yao, R.Z.; Li, Z.P.; Wang, X.W.; Dong, X.L.; Qu, H.; Li, Q.X.; Li, N.; Chi, H.; et al. Intuitive label-free sers detection of bacteria using aptamer-based in situ silver nanoparticles synthesis. *Anal. Chem.* **2017**, *89*, 9836–9842. [CrossRef] [PubMed]
135. Efrima, S.; Zeiri, L. Understanding sers of bacteria. *J. Raman Spectrosc.* **2009**, *40*, 277–288. [CrossRef]
136. Chauvet, R.; Lagarde, F.; Charrier, T.; Assaf, A.; Thouand, G.; Daniel, P. Microbiological identification by surface-enhanced raman spectroscopy. *Appl. Spectrosc. Rev.* **2017**, *52*, 123–144. [CrossRef]
137. Shan, S.J.; Liu, D.Z.; Liu, R.R.; Zhu, Y.Y.; Li, T.; Zhang, F.M.; An, L.G.; Yang, G.W.; Li, H. Non-mammalian toll-like receptor 18 (tlr18) recognizes bacterial pathogens in common carp (cyprinus carpio l.): Indications for a role of participation in the nf-kappa b signaling pathway. *Fish Shellfish Immunol.* **2018**, *72*, 187–198. [CrossRef] [PubMed]
138. Draz, M.S.; Lu, X.N. Development of a loop mediated isothermal amplification (lamp)—surface enhanced raman spectroscopy (sers) assay for the detection of salmonella enterica serotype enteritidis. *Theranostics* **2016**, *6*, 522–532. [CrossRef] [PubMed]
139. Gracie, K.; Correa, E.; Mabbott, S.; Dougan, J.A.; Graham, D.; Goodacre, R.; Faulds, K. Simultaneous detection and quantification of three bacterial meningitis pathogens by sers. *Chem. Sci.* **2014**, *5*, 1030–1040. [CrossRef]
140. Ma, Q.; Li, Y.L.; Gong, N.C.; Jiang, X.; Huan, S.Y. Surface enhanced raman spectroscopy sensor based on magnetic beads-induced nanoparticles aggregation for detection of bacterial deoxyribonucleic acid. *Chin. J. Anal. Chem.* **2015**, *43*, 1676–1681. [CrossRef]
141. Wang, Y.; Rauf, S.; Grewal, Y.S.; Spadafora, L.J.; Shiddiky, M.J.A.; Cangelosi, G.A.; Schlucker, S.; Trau, M. Duplex microfluidic sers detection of pathogen antigens with nanoyeast single-chain variable fragments. *Anal. Chem.* **2014**, *86*, 9930–9938. [CrossRef] [PubMed]
142. Carlson, H.K.; Iavarone, A.T.; Coates, J.D. Surfaceomics and surface-enhanced raman spectroscopy of environmental microbes: Matching cofactors with redox-active surface proteins. *Proteomics* **2013**, *13*, 2761–2765. [CrossRef] [PubMed]
143. Kang, H.; Yang, J.K.; Noh, M.S.; Jo, A.; Jeong, S.; Lee, M.; Lee, S.; Chang, H.; Lee, H.; Jeon, S.J.; et al. One-step synthesis of silver nanoshells with bumps for highly sensitive near-ir sers nanoprobes. *J. Mater. Chem. B* **2014**, *2*, 4415–4421. [CrossRef]
144. Xu, J.J.; Turner, J.W.; Idso, M.; Biryukov, S.V.; Rognstad, L.; Gong, H.; Trainer, V.L.; Wells, M.L.; Strom, M.S.; Yu, Q.M. In situ strain-level detection and identification of vibrio parahaemolyticus using surface-enhanced raman spectroscopy. *Anal. Chem.* **2013**, *85*, 2630–2637. [CrossRef] [PubMed]
145. Stephen, K.E.; Homrighausen, D.; DePalma, G.; Nakatsu, C.H.; Irudayaraj, J. Surface enhanced raman spectroscopy (sers) for the discrimination of arthrobacter strains based on variations in cell surface composition. *Analyst* **2012**, *137*, 4280–4286. [CrossRef] [PubMed]
146. Muhlig, A.; Bocklitz, T.; Labugger, I.; Dees, S.; Henk, S.; Richter, E.; Andres, S.; Merker, M.; Stockel, S.; Weber, K.; et al. Loc-sers: A promising closed system for the identification of mycobacteria. *Anal. Chem.* **2016**, *88*, 7998–8004. [CrossRef] [PubMed]
147. DeJong, C.S.; Wang, D.I.; Polyakov, A.; Rogacs, A.; Simske, S.J.; Shkolnikov, V. Bacterial detection and differentiation via direct volatile organic compound sensing with surface enhanced raman spectroscopy. *Chemistryselect* **2017**, *2*, 8431–8435. [CrossRef]

148. Guo, J.X.; Liu, Y.; Chen, Y.L.; Li, J.Q.; Ju, H.X. A multifunctional sers sticky note for real- time quorum sensing tracing and inactivation of bacterial biofilms. *Chem. Sci.* **2018**, *9*, 5906–5911. [CrossRef] [PubMed]
149. Premasiri, W.R.; Lee, J.C.; Sauer-Budge, A.; Theberge, R.; Costello, C.E.; Ziegler, L.D. The biochemical origins of the surface-enhanced raman spectra of bacteria: A metabolomics profiling by sers. *Anal. Bioanal. Chem.* **2016**, *408*, 4631–4647. [CrossRef] [PubMed]
150. Zukovskaja, O.; Jahn, I.J.; Weber, K.; Cialla-May, D.; Popp, J. Detection of pseudomonas aeruginosa metabolite pyocyanin in water and saliva by employing the sers technique. *Sensors* **2017**, *17*. [CrossRef] [PubMed]
151. Wang, F.; Yang, H.J.; He, H.B.; Wang, C.F.; Gao, Y.D.; Zhong, Q.F.; Wang, X.H.; Zeng, Y.J. Study on the hemolysin phenotype and the genetype distribution of staphyloccocus aureus caused bovine mastitis in shandong dairy farms. *Int. J. Appl. Res. Vet. Med.* **2011**, *9*, 416–421.
152. Wang, X.G.; Huang, J.M.; Feng, M.Y.; Ju, Z.H.; Wang, C.F.; Yang, G.W.; Yuan, J.D.; Zhong, J.F. Regulatory mutations in the a2m gene are involved in the mastitis susceptibility in dairy cows. *Anim. Genet.* **2014**, *45*, 28–37. [CrossRef] [PubMed]
153. Zhang, Z.J.; Han, X.M.; Wang, Z.M.; Yang, Z.; Zhang, W.M.; Li, J.; Yang, H.H.; Ling, X.Y.; Xing, B.G. A live bacteria sers platform for the in situ monitoring of nitric oxide release from a single mrsa. *Chem. Commun.* **2018**, *54*, 7022–7025. [CrossRef] [PubMed]
154. Pekdemir, M.E.; Erturkan, D.; Kulah, H.; Boyaci, I.H.; Ozgen, C.; Tamer, U. Ultrasensitive and selective homogeneous sandwich immunoassay detection by surface enhanced raman scattering (sers). *Analyst* **2012**, *137*, 4834–4840. [CrossRef] [PubMed]
155. He, C.Q.; Liu, Y.X.; Wang, H.M.; Hou, P.L.; He, H.B.; Ding, N.Z. New genetic mechanism, origin and population dynamic of bovine ephemeral fever virus. *Vet. Microbiol.* **2016**, *182*, 50–56. [CrossRef] [PubMed]
156. Hou, P.L.; Wang, H.M.; Zhao, G.M.; He, C.Q.; He, H.B. Rapid detection of infectious bovine rhinotracheitis virus using recombinase polymerase amplification assays. *BMC Vet. Res.* **2017**, *13*. [CrossRef] [PubMed]
157. Liang, J.W.; Tian, F.L.; Lan, Z.R.; Huang, B.; Zhuang, W.Z. Selection characterization on overlapping reading frame of multiple-protein-encoding p gene in newcastle disease virus. *Vet. Microbiol.* **2010**, *144*, 257–263. [CrossRef] [PubMed]
158. Hou, P.L.; Zhao, G.M.; He, C.Q.; Wang, H.M.; He, H.B. Biopanning of polypeptides binding to bovine ephemeral fever virus g(1) protein from phage display peptide library. *BMC Vet. Res.* **2018**, *14*. [CrossRef] [PubMed]
159. Yu, J.; Wu, J.Q.; Zhang, Y.Y.; Guo, L.H.; Cong, X.Y.; Du, Y.J.; Li, J.; Sun, W.B.; Shi, J.L.; Peng, J.; et al. Concurrent highly pathogenic porcine reproductive and respiratory syndrome virus infection accelerates haemophilus parasuis infection in conventional pigs. *Vet. Microbiol.* **2012**, *158*, 316–321. [CrossRef] [PubMed]
160. Lim, J.Y.; Nam, J.S.; Yang, S.E.; Shin, H.; Jang, Y.H.; Bae, G.U.; Kang, T.; Lim, K.I.; Choi, Y. Identification of newly emerging influenza viruses by surface-enhanced raman spectroscopy. *Anal. Chem.* **2015**, *87*, 11652–11659. [CrossRef] [PubMed]
161. Kaminska, A.; Witkowska, E.; Winkler, K.; Dziecielewski, I.; Weyher, J.L.; Waluk, J. Detection of hepatitis b virus antigen from human blood: Sers immunoassay in a microfluidic system. *Biosens. Bioelectron.* **2015**, *66*, 461–467. [CrossRef] [PubMed]
162. Sun, Y.; Xu, L.; Zhang, F.D.; Song, Z.G.; Hu, Y.W.; Ji, Y.J.; Shen, J.Y.; Li, B.; Lu, H.Z.; Yang, H.F. A promising magnetic sers immunosensor for sensitive detection of avian influenza virus. *Biosens. Bioelectron.* **2017**, *89*, 906–912. [CrossRef] [PubMed]
163. Zheng, S.; Shi, J.; Wu, X.; Peng, Z.; Xin, C.; Zhang, L.; Liu, Y.; Gao, M.; Xu, S.; Han, H.; et al. Presence of torque teno sus virus 1 and 2 in porcine circovirus 3-positive pigs. *Transbound. Emerg. Dis.* **2018**, *65*, 327–330. [CrossRef] [PubMed]
164. Zheng, S.; Wu, X.; Zhang, L.; Xin, C.; Liu, Y.; Shi, J.; Peng, Z.; Xu, S.; Fu, F.; Yu, J.; et al. The occurrence of porcine circovirus 3 without clinical infection signs in shandong province. *Transbound. Emerg. Dis.* **2017**, *64*, 1337–1341. [CrossRef] [PubMed]
165. Luo, Z.H.; Li, W.T.; Lu, D.L.; Chen, K.; He, Q.G.; Han, H.Y.; Zou, M.Q. A sers-based immunoassay for porcine circovirus type 2 using multi-branched gold nanoparticles. *Microchim. Acta* **2013**, *180*, 1501–1507. [CrossRef]
166. Luo, Z.H.; Chen, L.N.; Liang, C.J.; Wei, Q.M.; Chen, Y.; Wang, J. Porous carbon films decorated with silver nanoparticles as a sensitive sers substrate, and their application to virus identification. *Microchim. Acta* **2017**, *184*, 3505–3511. [CrossRef]

167. Reyes, M.; Piotrowski, M.; Ang, S.K.; Chan, J.Q.; He, S.A.; Chu, J.J.H.; Kah, J.C.Y. Exploiting the anti-aggregation of gold nanostars for rapid detection of hand, foot, and mouth disease causing enterovirus 71 using surface-enhanced raman spectroscopy. *Anal. Chem.* **2017**, *89*, 5373–5381. [CrossRef] [PubMed]
168. Sanchez-Purra, M.; Carre-Camps, M.; de Puig, H.; Bosch, I.; Gehrke, L.; Hamad-Schifferli, K. Surface-enhanced raman spectroscopy-based sandwich immunoassays for multiplexed detection of zika and dengue viral biomarkers. *ACS Infect. Dis.* **2017**, *3*, 767–776. [CrossRef] [PubMed]
169. Lin, D.; Gong, T.X.; Hong, Z.Y.; Qiu, S.F.; Pan, J.J.; Tseng, C.Y.; Feng, S.Y.; Chen, R.; Kong, K.V. Metal carbonyls for the biointerference-free ratiometric surface-enhanced raman spectroscopy-based assay for cell-free circulating DNA of epstein-barr virus in blood. *Anal. Chem.* **2018**, *90*, 7139–7147. [CrossRef] [PubMed]
170. Harpster, M.H.; Zhang, H.; Sankara-Warrier, A.K.; Ray, B.H.; Ward, T.R.; Kollmar, J.P.; Carron, K.T.; Mecham, J.O.; Corcoran, R.C.; Wilson, W.C.; et al. Sers detection of indirect viral DNA capture using colloidal gold and methylene blue as a raman label. *Biosens. Bioelectron.* **2009**, *25*, 674–681. [CrossRef] [PubMed]
171. Malvadkar, N.A.; Demirel, G.; Poss, M.; Javed, A.; Dressick, W.J.; Demirel, M.C. Fabrication and use of electroless plated polymer surface-enhanced raman spectroscopy substrates for viral gene detection. *J. Phys. Chem. C* **2010**, *114*, 10730–10738. [CrossRef]
172. Huang, Y.H.; Wang, X.J.; Zhang, F.; Huo, X.B.; Fu, R.S.; Liu, J.J.; Sun, W.B.; Kang, D.M.; Jing, X. The identification of a bacterial strain bgi-1 isolated from the intestinal flora of blattella germanica, and its anti-entomopathogenic fungi activity. *J. Econ. Entomol.* **2013**, *106*, 43–49. [CrossRef] [PubMed]
173. Dina, N.E.; Gherman, A.M.R.; Chis, V.; Sarbu, C.; Wieser, A.; Bauer, D.; Haisch, C. Characterization of clinically relevant fungi via sers fingerprinting assisted by novel chemometric models. *Anal. Chem.* **2018**, *90*, 2484–2492. [CrossRef] [PubMed]
174. Witkowska, E.; Jagielski, T.; Kaminska, A.; Kowalska, A.; Hryncewicz-Gwozdz, A.; Waluk, J. Detection and identification of human fungal pathogens using surface-enhanced raman spectroscopy and principal component analysis. *Anal. Methods* **2016**, *8*, 8427–8434. [CrossRef]
175. Zivanovic, V.; Semini, G.; Laue, M.; Drescher, D.; Aebischer, T.; Kneipp, J. Chemical mapping of leishmania infection in live cells by sers microscopy. *Anal. Chem.* **2018**, *90*, 8154–8161. [CrossRef] [PubMed]
176. Krause, D.C.; Hennigan, S.L.; Henderson, K.C.; Clark, H.J.; Dluhy, R.A. Variable selection and biomarker correlation in the analysis of mycoplasma pneumoniae strains by surface-enhanced raman spectroscopy. *Anal. Lett.* **2017**, *50*, 2412–2425. [CrossRef]
177. Garrett, N.L.; Sekine, R.; Dixon, M.W.A.; Tilley, L.; Bambery, K.R.; Wood, B.R. Bio-sensing with butterfly wings: Naturally occurring nano-structures for sers-based malaria parasite detection. *Phys. Chem. Chem. Phys.* **2015**, *17*, 21164–21168. [CrossRef] [PubMed]

*nanomaterials*

MDPI

*Review*

# Detection and Characterization of Antibiotic-Resistant Bacteria Using Surface-Enhanced Raman Spectroscopy

**Kaidi Wang [1], Shenmiao Li [1], Marlen Petersen [1], Shuo Wang [2] and Xiaonan Lu [1,*]**

[1] Food, Nutrition and Health Program, Faculty of Land and Food Systems, The University of British Columbia, Vancouver, BC V6T1Z4, Canada; kaidi.wang@ubc.ca (K.W.); shenmiao.ivy.li@gmail.com (S.L.); Marlen.Petersen@web.de (M.P.)

[2] Tianjin Key Laboratory of Food Science and Health, School of Medicine, Nankai University, Tianjin 300371, China; wangshuo@nankai.edu.cn

* Correspondence: xiaonan.lu@ubc.ca; Tel.: +1-604-822-2551

Received: 25 August 2018; Accepted: 23 September 2018; Published: 26 September 2018

**Abstract:** This mini-review summarizes the most recent progress concerning the use of surface-enhanced Raman spectroscopy (SERS) for the detection and characterization of antibiotic-resistant bacteria. We first discussed the design and synthesis of various types of nanomaterials that can be used as the SERS-active substrates for biosensing trace levels of antibiotic-resistant bacteria. We then reviewed the tandem-SERS strategy of integrating a separation element/platform with SERS sensing to achieve the detection of antibiotic-resistant bacteria in the environmental, agri-food, and clinical samples. Finally, we demonstrated the application of using SERS to investigate bacterial antibiotic resistance and susceptibility as well as the working mechanism of antibiotics based on spectral fingerprinting of the whole cells.

**Keywords:** SERS; chemometrics; resistance; biosensing; rapid detection

---

## 1. Introduction

Detection of pathogenic and spoilage bacteria is still a major concern to clinical, agri-food, and environmental agencies and laboratories [1]. The leading challenge is the detection speed [1]. Since the contamination level of bacteria may be relatively low and the sample matrices can significantly influence accurate and reproducible detection, extensive sample preparation steps are always required to separate the targeted bacteria from the sample matrices along with pre-enrichment [2,3]. Because the detection includes all the times starting from obtaining the samples to the signal readout, both separation and bacterial enrichment account for most of the times for bacterial detection rather than the final real detection using an instrument or a sensor [4]. For example, the conventional plating assay will take several days to confirm the growth of the targeted bacterial colony [5]. In comparison, molecular-based detection methods, such as polymerase chain reaction (PCR), requires relatively less time than the plating assay but still cannot fully avoid separation and bacterial pre-enrichment [6]. Recently, matrix assisted laser desorption ionization time-of-flight (MALDI-TOF) spectrometry has attracted considerable interest for the rapid identification of pathogens by profiling bacterial proteins from the whole cells [7]. However, this method is not suitable for characterizing a mixed sample [8] and still requires the priori cultivation and sample preparation procedure [9]. An alternative method is surface-enhanced Raman spectroscopy (SERS), an advanced Raman spectroscopic technique that enhances the vibrational modes of molecules adsorbed on or in the vicinity to the surface of metal nanoparticles. SERS provides rapid, ultra-sensitive and accurate detection with minimum requirement for sample handling and preparation.

Antibiotic resistance of pathogenic bacteria is still a leading concern to clinics as well as agri-food and veterinary medicine [10]. The key battle is to perform an accurate diagnosis of the pattern of bacterial antibiotic resistance in an early manner. Otherwise, only the broad-spectrum antibiotics can be used to treat this type of bacterial infections [11]. As aforementioned, the conventional microbiological testing, such as the determination of minimum inhibitory concentration (MIC) using the broth microdilution method, is highly time-consuming. Besides, PCR-based testing of the targeted antibiotic-resistant genes requires highly trained personnel and has a potential risk of cross contamination [12,13]. Another major limitation of this approach is that the presence of the resistance genes may not necessarily confer to the clinically relevant phenotypic resistance of bacteria [14]. Microarray offers the ability to detect a broad range of resistance genes present in the bacterial isolates with high sensitivity and specificity. However, similar to the PCR-based method, results obtained from microarrays may not always correlate to the phenotypic resistance [14]. Although MALDI-TOF mass spectroscopy can potentially differentiate the resistant and susceptible isolates based on the spectral features [7], it requires additional chemicals as the matrix for the performance of MALDI [14]. Alternative technology that can detect and characterize bacterial antibiotic resistance is therefore highly required. SERS is a powerful biochemical fingerprinting technique as it can accurately reflect the macromolecular profiles and changes that occur within the bacterial cells due to the action of the antibiotics [15].

In this mini-review paper, we will evaluate the use of SERS coupled with chemometrics as a tool to detect the trace level of antibiotic-resistant bacteria and characterize the mechanism of bacterial antibiotic resistance in an ultra-fast manner. The recent progress in this research area will be summarized and discussed mainly focusing on the following three perspectives: (1) the nanomaterials that can be used as the SERS substrates for sensing a low concentration of bacterial cells; (2) tandem-SERS technology to detect antibiotic-resistant bacteria in a sample matrix; and (3) characterizing the mechanism of bacterial antibiotic resistance and susceptibility using SERS and chemometrics.

## 2. Surface-Enhanced Raman Spectroscopy (SERS) for Sensing Trace Level of Bacteria

### 2.1. Mechanism of SERS

SERS is a derivative of Raman spectroscopy with the aid of nanomaterials. Numerous research studies have been conducted during the past four decades about using SERS for trace detection of the targeted analytes [16–23]. Different from the conventional Raman spectroscopic technology, SERS signal can be significantly enhanced due to both electromagnetic enhancement and chemical enhancement, with the former being the dominant contributor [24]. Electromagnetic enhancement is generated from the localized surface plasmon resonance (LSPR) in the vicinity of the nanostructured surface of noble metals, such as silver and gold [25,26]. Highly localized regions of amplified electromagnetic fields caused by LSPR are called "hot spots", which usually occurs in the gaps, crevices, or sharp vertices of supporting plasmonic materials (Figure 1a). In comparison, chemical enhancement is due to the electron transfer between the analyte molecule and the surface of the nanostructure when the energy of the incident light matches the electron transfer energy (Figure 1b) [27]. This will lead to the change of molecular polarization and subsequently enhance the Raman signal approximately 100 times. Theoretically, total SERS enhancement factors may approach to ~$10^{14}$ depending on the nanomaterials used. For additional details, the authors are encouraged to refer to serial publications from the Van Duyne research group [27–30].

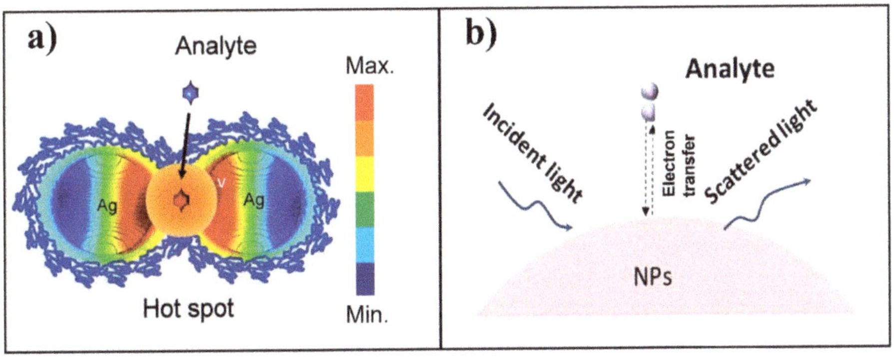


**Figure 1.** Two mechanisms contributed to surface-enhanced Raman spectroscopy (SERS). (**a**) Electromagnetic enhancement of SERS-active silver nanoparticles. SERS "hot-spot" is generated in the gap between two close nanoparticles. (**b**) Chemical enhancement resulting from electron transfer between analytes and the surface of nanoparticles. Reproduced with permission [31]. Copyright Royal Society of Chemistry, 2014. Reproduced with permission [32]. Copyright Elsevier B.V., 2017.

### 2.2. SERS-Active Substrates for Bacterial Detection

Because SERS can reach to single molecule detection, it has been widely applied for the detection of various analytes in an ultra-fast manner (e.g., a few seconds to less than a minute). In general, the reproducibility of the SERS signal is getting worse along with the increase of the size of the analyte [33]. For example, it is extremely challenging to harvest a reproducible SERS signal for a bacterial cell than that of a small chemical molecule, such as antibiotics and pesticides [34]. Although successful discrimination of bacteria by using SERS was reported by different research groups [15,35,36], the real world application is still extremely challenging, such as the low concentration of the targeted bacteria in the sample and a relatively large amount of interference sample components. Therefore, researchers have been developing various types of SERS-active substrates to enhance the signal intensity as well as generate more reproducible SERS signals for different biological samples, such as bacteria and viruses. Both "top-down" and "bottom-up" methods have been used for the synthesis of SERS-active substrates [37]. For the "top-down" method, large multi-dimensional materials are reduced to ideal nanoscale structures using direct fabrication process [38]. In comparison, the "bottom-up" method refers to the development of complex nanoscale structures from simple molecules or atoms [39].

#### 2.2.1. Direct SERS

Generally, SERS-active nanostructures are composed of two types of substrates: solid surface-based substrates and colloidal substrates. The solid surface-based substrates can accurately control the formation of "hot spots". Once the bacteria cells are closed to the "hot-spot" on the surface of the solid substrate, a significant SERS effect will be achieved. For example, a recent study presented a label-free SERS-based method to detect and identify *Salmonella enterica* and *Escherichia coli* adsorbed on the silver dendrites [40]. Since the nanoparticles were already closely aligned on the stem and branches, "hot spots" could be generated without any aggregation process. This also contributed to producing uniform and homogenous sample spots after drying, which eliminated the spot-to-spot variation of the collected SERS signals. SERS spectra collected using the silver dendrites were consistent and robust enough for the detection and identification of bacteria with a limit of detection (LOD) as low as $10^4$ colony-forming unit (CFU) per mL. Besides, porous anodic aluminum oxide (AAO) has been widely used as the substrate for the synthesis of functional nanostructures by coating a thin layer of gold or silver to develop a nanostructured noble metal substrate to enhance SERS signal intensity [41]. Ji and co-authors reported a three-dimensional nanostructure fabricated by depositing silver NPs into AAO templates using a simple electrochemical deposition method [42], demonstrating well-ordered micro/nanostructures when it was characterized by field emission scanning electron microscopy. The homogeneity of SERS substrates is the key to the reproducibility of SERS spectra and even minor variation in the surface morphology can result in significant changes in the enhancement. Due to the well-organized structure of decorated AAO membranes, the distribution of "hot-spots" is uniform, which can eventually improve the SERS spectral reproducibility [43].

In addition, various colloid systems of gold or silver have been synthesized as the liquid format of SERS substrates for the detection of bacterial cells [44]. A more uniform distribution of noble metal nanoparticles on the surface of bacterial cells can be achieved to improve the SERS spectral reproducibility compared to that by using the solid SERS substrates [45]. A SERS application employing a synthesis of silver nanocolloids coating on a bacterial cell wall can detect the live bacteria in drinking water down to $2.5 \times 10^2$ CFU/mL [46]. Another study conducted by Chen and colleagues applied Ag colloids for the discrimination of *E. coli*, *Pseudomonas aeruginosa*, methicillin-resistant *Staphylococcus aureus* (MRSA) and *Listeria*. In situ synthesis of Ag nanoparticles and the addition of Triton X-100 significantly improved the sensitivity of SERS detection [47]. A simple method of preparing SERS substrates was described by filtering Ag or Au colloidal particles onto a ceramic filter, onto which the bacterial suspensions were then filtered [48]. This method allowed the homogeneous distribution of bacteria on the surface of the substrate, which increased the sensitivity of SERS detection. A microfluidic "lab-on-a-chip" platform can be used to further improve the reproducibility of SERS signal by mixing the silver/gold nanocolloids with bacterial cells in a controlled fluidic manner with limited precipitation of individual nanoparticles on the substrate, in which case the channel in the microfluidic device could avoid spectral interference and enhance the sensitivity of bacterial detection [49]. SERS-microfluidic systems have been used to classify multiple foodborne pathogens using chemometrics and quantify single pathogenic bacterial cells. For example, Mungroo and others successfully distinguished eight foodborne pathogenic bacterial species using microfluidic-integrated SERS substrate and chemometrics, including principal component analysis (PCA) and linear discriminant analysis (LDA) [50]. A SERS-based microfluidic system was developed for the discrimination of *E. coli* strains with the spectral recording time reduced to 1 s [51]. Ag nanoparticles were injected into the bacterial suspension to facilitate the aggregation of nanocolloids on the bacterial cells. Besides, a SERS substrate composed of 3D Ag@ZnO nanostructures was also integrated into a microfluidic device for SERS fingerprinting detection of a single living cell [52]. Colloidal substrate seems to be more popular due to its simple and cost-effective fabrication, but solid surface-based substrates are more favorable for the detection of water-insoluble substances [53]. A variety of SERS nanomaterials used for bacterial biosensing have been summarized in Table 1.

**Table 1.** Summary of SERS-active nanomaterials used for the detection of bacteria.

| SERS-Active Nanomaterial | Target Bacteria | LOD (CFU/mL) | LOQ (CFU/mL) | Sample Matrix | Detection Time | Chemometric Models | COMMENTS | Ref. |
|---|---|---|---|---|---|---|---|---|
| AgNPs | *E. coli* | $8.0 \times 10^2$ | N/A | N/A | 3.1 h | - | Direct, microfluidic | [51] |
| AgNPs | methicillin-resistant *S. aureus* (MRSA) | N/A | N/A | N/A | 3.3 min | DFA, HCA | Direct, microfluidic concentration | [54] |
| AgNPs | *M. tuberculosis* | - | N/A | - | 1 h | PCA, LDA | Direct, microfluidic concentration | [55] |
| AuNP surface | *K. pneumoniae* | N/A | N/A | N/A | 30 min | PCA | Direct, fluoroquinolone-resistant | [56] |
| AgNPs | *E. coli, A. calcoaceticus, B. megaterium, P. aeruginosa* | N/A | N/A | N/A | N/A | N/A | Direct | [57] |
| AgNPs | *E. coli, S. cohnii* | N/A | N/A | N/A | 10 s | N/A | Direct, convective assembly | [58] |
| AgNPs and AuNPs | *E. coli, S. cohnii* | N/A | N/A | N/A | N/A | N/A | Direct, layer-by-layer deposition | [45] |
| AgNPs | *E. coli, S. epidermidis* | $2.5 \times 10^2$ | N/A | N/A | 10 min | HCA | Direct, in situ adsorption | [46] |
| AgNPs | *E. coli, M. morganii, E. lactis, L. casei* | NA | N/A | N/A | <5 min | PCA | Direct, in situ synthesis | [59] |
| Ag nanospheres | *E. coli, S. typhimurium, S. aureus* | 10 | N/A | N/A | N/A | CVA | Direct, self-assembly, Ag nanocrystals | [60] |
| Ag nanorods | *A. baumannii, E. coli, K. pneumoniae, P. aeruginosa, S. aureus* | N/A | N/A | N/A | N/A | PCA, HCA, PLS-DA | Indirect, vancomycin-coated | [61] |
| Octupolar metastructures | *Brucella* | $10^4$ | N/A | N/A | N/A | N/A | Indirect, bacteriophage, EBL fabrication | - |
| Au nanorods | *E. coli* | $3.5 \times 10^1$ | $3.5 \times 10^2$ | N/A | <2 h | N/A | Indirect, Raman reporter, biotin-avidin, magnetic core | [62] |
| Ag nanocubes | *E. coli* | $10^2$ | N/A | N/A | - | N/A | Indirect, Raman reporter, polyclonal antibody | [63] |
| AgNPs, AuNPs, and Ag/Au core shell NP | *E. coli* O157:H7, *S.* Typhimurium, *S. aureus* | $10^2$–$10^3$ | N/A | N/A | <30 min | N/A | Indirect, Raman reporter, aptamers, multiplex detection | [64] |
| Au "nanopopcorn" @ single wall carbon nanotubes | *E. coli* | $10^2$ | $10^2$ | N/A | - | N/A | Indirect, antibody, photothermal inactivation | [65] |
| AuNP @ graphene oxide | MRSA | 5 | N/A | N/A | - | N/A | Indirect, Raman reporter, photothermal inactivation | [66] |
| Au "nanoovals" | *E. coli* | $2.1 \times 10^2$ | N/A | Chicken broth, apple juice, soil solution | 50 s | N/A | Indirect, Raman reporter, antibody, DEP concentration | [67] |
| AuNPs | *Mycobacterium avium* subsp. *Paratuberculosis* | $5.0 \times 10^2$ | N/A | Milk | <24 h | N/A | Indirect, Raman reporter, antibody | [68] |

**Table 1.** *Cont.*

| SERS-Active Nanomaterial | Target Bacteria | LOD (CFU/mL) | LOQ (CFU/mL) | Sample Matrix | Detection Time | Chemometric Models | COMMENTS | Ref. |
|---|---|---|---|---|---|---|---|---|
| Au "nanopopcorn" @ graphene oxide | MRSA | 10 | N/A | N/A | - | N/A | Indirect, Raman reporter, aptamer | - |
| Ag nanorod arrays | *S.* Enteritidis, *S. enterica* | $10^2$ | N/A | Mung bean sprouts samples | - | PCA, PLS-DA | Indirect, vancomycin-coated surface | [69] |
| C | *S.* Typhimurium, *S. aureus* | $10^3$ | N/A | Spinach | N/A | N/A | Indirect, antibody, $Fe_3O_4/SiO_2$ secondary NPs | [70] |
| $Ag/SiO_2$ core/shell NPs | *S.* Typhimurium | $10^8$ | N/A | N/A | N/A | N/A | Indirect, Raman reporter, antibody | [71] |
| Au "nanopopcorn" | *S.* Typhimurium DT 104 | 10 | N/A | Romaine lettuce | 5 min | N/A | Indirect, Raman reporter, monoclonal antibody | [72] |
| $SiO_2$/Au and Au/Ag core/shell NPs | *S.* Typhimurium | 15 | 15 | Milk | N/A | N/A | Indirect, Raman reporters, aptamers | [73] |
| Au/Ag core–shell nanoparticles | *V. parahaemolyticus* | 10 | 10 | N/A | N/A | N/A | Indirect, Raman reporters, aptamers | [74] |
| Au nanopopcorn | *S.* Typhimurium DT 104 | 10 | N/A | N/A | N/A | N/A | Indirect, Raman reporter, antibody, photothermal inactivation | [75] |
| $Fe_3O_4$/Au core/shell NPs | *S.* Typhimurium, *S. aureus* | 15 | $10^2$ | Pork sample | N/A | N/A | Indirect, aptamer, magnetic separation | [76] |
| $MnFe_2O_4$/Au core/shell | *S. aureus* | 10 | N/A | Apple, pear, and grapes peels | N/A | N/A | Indirect, Raman reporter, aptamer, magnetic separation | [77] |
| Au nanoaggregate-embedded beads | *S. aureus* | N/A | N/A | N/A | N/A | N/A | Indirect, Raman reporter, antibody | [78] |
| AgNPs | *S. aureus* | 15 | 15 | Urine, blood, or pleural and ascites fluids | N/A | N/A | Direct, antibody, aptamer, Raman reporter | [79] |
| $Fe_3O_4$/Au core/shell NP | *S. aureus* | 1 | N/A | N/A | N/A | N/A | Indirect, antibody, magnetic concentration/separation | [80] |
| Au/Ag core/shell nanorod arrays | *S. xylosus*, *L. monocytogenes*, *E. faecium* | 50 | N/A | N/A | PCA | N/A | Indirect, Raman reporter | [81] |

SERS has been widely applied for the differentiation of antibiotic-resistant strain and antibiotic-sensitive strains possibly due to the variation in the biochemical compositions of bacterial cell membrane and cell wall. In a recent study, Li and others reported that surface-enhanced resonance Raman spectroscopy (SERRS) could achieve almost a 100% accuracy for the differentiation between carbapenem-resistant *E. coli* and carbapenem-sensitive *E. coli* [82]. Lu and coauthors developed a microfluidic SERS platform for a successful high-throughput screening and differentiation between MRSA and methicillin-sensitive *Staphylococcus aureus* (MSSA). In addition, the SERS characterization of bacterial phenotypic profiles had a good correlation to the multilocus sequence typing as well as antibiotic characterization using PCR, demonstrating the possibility of applying SERS as the alternative to detect antibiotic resistance and track the outbreak of pathogenic bacteria [54]. In another study, Mühlig and coauthors applied a similar SERS microfluidic chip for the differentiation of various species of mycobacteria, including both nontuberculous mycobacteria and *Mycobacterium tuberculosis* complex [55].

2.2.2. Indirect SERS

The aforementioned SERS substrates are related to "direct sensing" of the analyte (e.g., a bacterium) by using a laser with the wavenumbers of mainly 532, 633, and 785 nm [53]. In other words, the collected SERS spectral features are directly associated with the chemical compositions of the targeted bacteria (Figure 2a). In comparison, SERS tags have been designed and used for "indirect sensing" of the analyte(s) (Figure 2b).

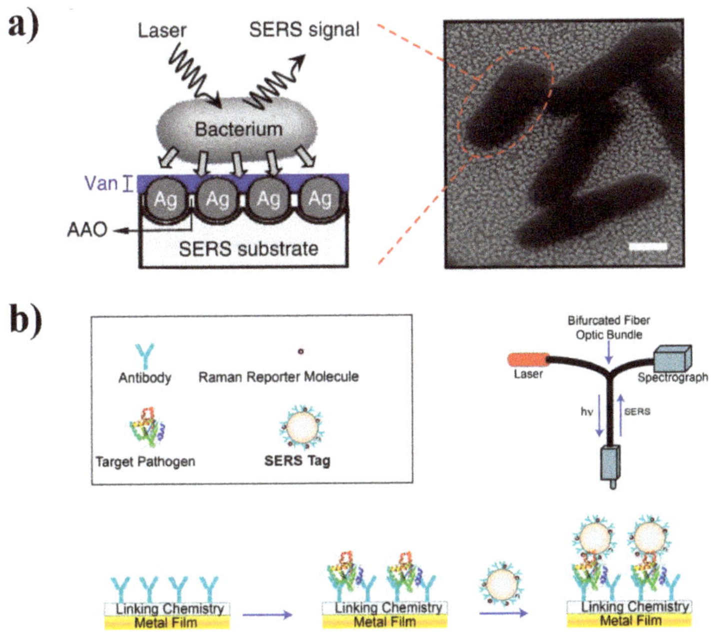


**Figure 2.** Representative "direct" (**a**) and "indirect" (**b**) SERS detection of bacteria. (**a**) Schematic diagram showing the SERS signal was directly collected from the bacterium on a vancomycin-coated Ag/AAO SERS-active substrate (left). Scanning electron microscope (SEM) image of bacteria on the substrate (scale bar, 500 nm) (right). (**b**) Schematic illustration of a sandwich-like indirect antibody-SERS detection. Key steps including: immobilization of antibody on the surface of metal substrate; capture of target bacteria by modified surface and labeling the target bacteria with SERS tag for detection. Reproduced with permission [83]. Copyright Springer Nature, 2011. Reproduced with permission [72]. Copyright Royal Society of Chemistry, 2011.

The schematic illustration of the SERS tag is shown in Figure 3. Specifically, a SERS-active molecule, such as rhodamine 6G, will be used as the tag molecule for the synthesis of a gold/silver nanostructure [72]. By conjugating with a separation element, such as an antibody, aptamer, or a molecularly-imprinted polymer, a functional SERS tag will be developed. This SERS tag can specifically recognize and capture the targeted analyte (e.g., a bacterium) from a complicated sample matrix to achieve separation and possibly enrichment as well [32].

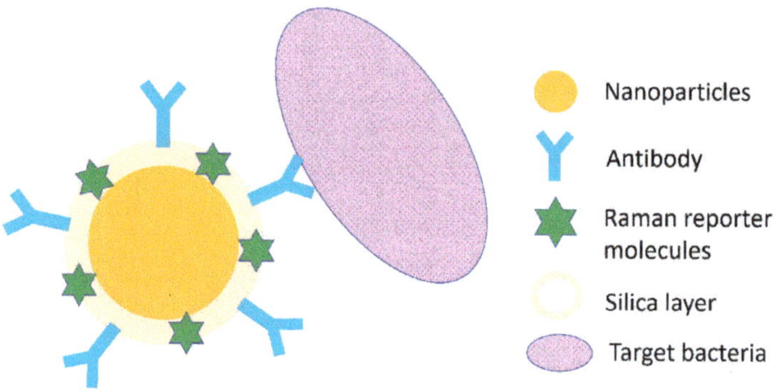


**Figure 3.** Schematic illustration of SERS tags.

Most indirect approaches use a sandwich-like immunosorbent assay format, which is similar to enzyme-linked immunosorbent assay (ELISA) [68]. The schematic illustration in Figure 2b shows the basic steps for developing a representative sandwich-structured indirect antibody-SERS method. Firstly, capturing antibodies are immobilized on the surface of a metal substrate. The second step is to capture the targeted pathogen from the sample matrix using these immobilized antibodies. Finally, the SERS tag will be introduced to label the targeted pathogen for Raman signal collection. The availability of the collected SERS signal is derived from the SERS tag molecule, but can indirectly indicate the availability and the concentration of the targeted bacteria in the sample matrix. This indirect SERS-tag technology is extremely useful for the detection of bacteria in a complicated sample matrix, such as a food, because the aforementioned direct SERS detection can be significantly affected by the food sample matrix if the sample pre-treatment is not fully complete [37]. For example, Duan and co-authors reported an indirect SERS-based method for the quantification of *S.* Typhimurium in milk (Figure 4a) [73]. *S.* Typhimurium interacted with $Fe_3O_4$/Au core/shell nanoparticles functionalized with specific aptamers and Raman reporters in conjugation to the same aptamer to form a sandwich-like complex. A linear correlation for bacteria concentration of ~10–$10^6$ CFU/mL and a low LOD of 15 CFU/mL were obtained in this study. *Vibrio parahaemolyticus* was successfully detected in shrimp and water samples using a similar approach [74]. The specific aptamer immobilized on the $SiO_2$-core-Au-shell nanoparticles was used to selectively capture *V. parahaemolyticus*, leading to a LOD of 10 CFU/mL. In another study, silver nanoparticles functionalized with antibodies and Raman reporter to serve as the SERS tags were successfully applied for rapid detection of *E. coli* to a concentration as low as $10^2$ CFU/mL [63]. Although several publications demonstrated a good separation capability and spectral reproducibility by integrating silver/gold nanoparticles with magnetic materials [84–86], we still believe a functional SERS tag with separation element is more effective at the current stage. More precise control of the numbers and orientations of the molecules on the surfaces of the magnetic nanoparticles have to be achieved [84]. In addition, a few studies reported the development of functional SERS tags by integrating both separation elements and magnetic beads to achieve an even better separation, enrichment, and signal enhancement capability [62,70,80,87]. For example, an LOD of 35 CFU/mL and LOQ of $3.5 \times 10^2$ CFU/mL for *E. coli* was reported using a combination of antibody-modified magnetic nanoparticles and gold nanorods labeled with the same antibodies in a sandwich-format detection strategy [62]. Besides, a recent study conducted by Kearns and colleagues reported a novel assay of using lectin-functionalized magnetic nanoparticles along with SERS-active nano-substrates functionalized with various antibodies to successfully capture and detect multiple antibiotic-resistant pathogens, including *Salmonella*, *E. coli* and MRSA at the single cell level in a simultaneous manner [88].

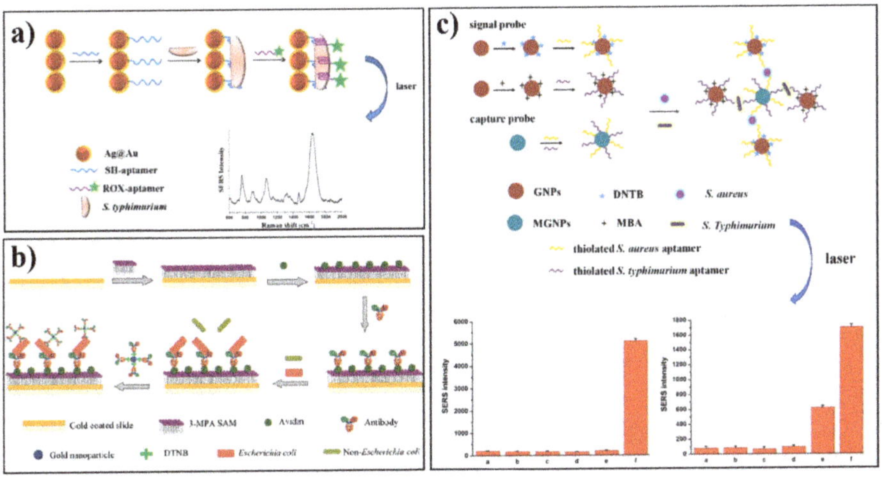


**Figure 4.** (**a**) Schematic illustration of aptamer-based SERS approach for the detection of *Salmonella* Typhimurium. Ag/Au core/shell nanoparticle was conjugated with a specific aptamer. The Raman reporter, X-rhodamine (ROX), was labeled on the same aptamer sequence. Nanoparticle-aptamer-target-aptamer-Raman reporter complexes enabled SERS detection. (**b**) Schematic illustration of the antibody-based sandwich-type SERS immunoassay for *Escherichia coli* enumeration. SERS tags were constructed by gold nanoparticles first coated with a Raman reporter molecule, 5,5′-dithiobis (2-nitrobenzoic acid) (DTNB), and subsequently with a corresponding antibody. (**c**) Multiplex detection of *Salmonella* Typhimurium and *Staphylococcus aureus* using aptamer-SERS immunoassay. $Fe_3O_4$ magnetic gold nanoparticles were labeled with unique Raman reporters and aptamers against *S. aureus* and *S.* Typhimurium and then employed into a sandwich-like assay. Reproduced with permission [73]. Copyright Elsevier B.V., 2015. Reproduced with permission [89]. Copyright Springer-Verlag, 2010. Reproduced with permission [76]. Copyright Elsevier B.V., 2015.

## 3. Tandem-SERS for Sensing Bacteria in a Sample Matrix

### 3.1. Tandem-SERS Methods

Tandem-SERS refers to conjugating the separation element to the SERS system that can achieve separation and detection simultaneously [90]. The aforementioned functional SERS tag with a separation element (e.g., antibody, aptamer, molecularly-imprinted polymer) is a classical tandem-SERS system. Due to the size of bacterial cells, a sandwich tandem-SERS structure is always developed [15,68] and the detailed illustration is shown in Figure 2b. Antibody is widely used as the recognition element due to its specificity to bacteria via a covalently-bound effect. An antibody conjugated with different SERS nanoprobes such as Ag@silica core-shell nanoparticles [71], popcorn-shaped Au nanoparticles [72], and single walled carbon nanotubes-Au nanoparticles [91] was used to detect normal *Salmonella* or multi-drug-resistant *Salmonella*. High correlation coefficients and LOD of 4 and 5 CFU $mL^{-1}$ were obtained using an antibody-SERS employing AuNPs via a sandwich immunoassay for detecting and enumerating *E. coli* (Figure 4b) [89]. The results of testing bacteria in lake and tap water samples were highly consistent with that of the classical plating assay.

Aptamer is another element that can be used and conjugated in tandem-SERS for the recognition, separation, and enrichment of specific bacterial pathogens. Aptamer-based SERS assay was able to monitor photothermal activity response of MRSA and multi-drug-resistant *Salmonella* DT104 through the change of Raman signal intensity of R6G [32]. Zhang and coauthors reported a simultaneous detection of *S.* Typhimurium and *S. aureus* using Au NPs-aptamer based SERS biosensor (Figure 4c). A high sensitivity with LOD of 35 and 15 CFU/mL for *S. aureus* and *S.* Typhimurium was achieved, respectively [76]. Another format of tandem-SERS was to include SERS sensing in a microfluidic device. A complicated design of the microfluidic device can realize the function of separation of bacterial cells from the sample matrix mainly [49]. Dielectrophoresis is an effective method for concentrating and trapping various types of nanoscale/microscale particles in a microfluidic device, including microorganisms [92]. It is also feasible to conjugate the aforementioned separation elements, such as aptamer, onto the microchannels to form a more comprehensive and effective tandem-SERS platform for simultaneous separation and detection [67]. Lin and co-authors developed a fast single-step SERS detection of *E. coli* O157:H7 at single cell level with speciation capability to sub-species. This was achieved by a multiplexing dual recognition SERS platform that combined specific antibody conjugated SERS tags with a microfluidic dielectrophoresis (Figure 5) [93].

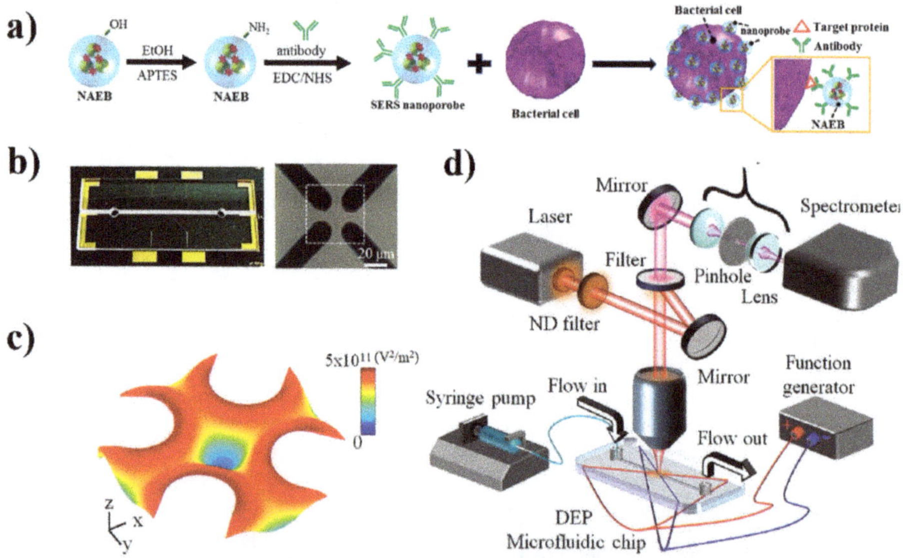


**Figure 5.** The integration of SERS nanoprobes and a microfluidic dielectrophoresis (DEP) device for rapid detection of single bacterium. (**a**) Schematic presentation of using antibody-conjugated nanoaggregate-embedded beads (NAEBs) as SERS nanoprobes for specific detection of bacteria. (**b**) Photograph of the microfluidic DEP device and close-up view of central capturing area with four the quadrupole electrodes. (**c**) The distribution of electric field of four microelectrodes in the microchannel. (**d**) Schematic illustration of the DEP-SERS configuration. Reproduced with permission [93]. Copyright WILEY-VCH Verlag GmbH & Co. KGaA, Weinheim, 2014.

### 3.2. Tandem-SERS Integrated with Multiple Capabilities

Another major application advantage of such a tandem-SERS platform is to enrich the bacterial cells and subsequently improve the detection sensitivity. Although SERS can theoretically detect a single molecule/cell, its real world application can only detect ~$10^3$ CFU/mL of bacteria, mainly due to the interference from the sample matrix components [94]. Therefore, a relatively large amount of samples therefore is required for the production of a meaningful SERS signal readout. In a recent study reported by Zhang and others, the SERS-active substrate composed of gold nanoparticles was integrated into the microfluidic device for rapid concentration and detection of *S. aureus* in liquid samples [95]. The SERS signal intensity of *S. aureus* after concentration in this device was over 100-fold compared to the signal obtained from the raw sample, leading to a LOD of $2 \times 10^2$–$2 \times 10^4$ CFU/mL. Hou and colleagues demonstrated a microfluidic system based on a discharge driven vortex technique to concentrate a bacterial suspension of *E. coli* F-amp and *Bacillus subtilis* for SERS detection. The combination of SERS and microfluidic with immunoassay techniques was able to selectively capture the targeted bacterial cells [96]. A SERS-based sandwich immunoassay employing antibody-coated magnetic nanoparticles for *E. coli* enumeration was also reported [97]. The authors accomplished a LOD of 8 CFU/mL by combining bacterial separation with SERS detection using specific SERS labels.

Combination of SERS platform and a filter (e.g., polymer fiber) has been recently used for the identification and detection of bacteria from clinical and environmental samples. For instance, Lin and others demonstrated a filter-like SERS substrate prepared with AuNPs embedded in mesoporous silica for the detection of *Staphylococcus aureus* from the aqueous samples [98]. The targeted cells could be concentrated on the filter-like substrates within a few seconds. Strong SERS signals with good bacterial discrimination were obtained without any need for pre-labeling, and the reproducibility was also significantly improved. More recently, Kamińska and colleagues presented a new label-free tandem-SERS platform for rapid detection of *Neisseria meningitidis* [99]. This bacterium is a Gram-negative diplococcus and one of the three major bacteria that cause acute bacterial meningitis. The applied SERS substrate was based on Si/ZnO layers and electrospun polymer mats covered with a thin layer of sputtered gold. A wide range of pore sizes makes the polymer mat an excellent material to filter bacteria from fluids and then immobilize them onto the SERS nanostructures for the collection of Raman signals, enabling the detection of single bacterial cells of *N. meningitidis* present in cerebrospinal fluid samples. A similar approach was developed to detect bacteria from blood plasma [100]. Covering the forcespun polymer mat with Au/Ag alloy turns it into a SERS-active platform, which can be used as a filter to separate the microorganisms from fluids and immobilize them on the surface of the mat during the measurement. *S. aureus*, *Pseudomonas aeruginosa*, and *S.* Typhimurium were successfully detected and identified from blood plasma using the developed platforms. These SERS-active nanostructures based on polymer mats provide the possibility for simultaneous filtration, immobilization, and enhancement of Raman signals in a few seconds, demonstrating a simple and low-cost method to analyze bacterial suspensions in biological fluids with SERS.

In addition, the tandem-SERS platform can achieve multiplex detection of bacteria by integrating several different elements into a single system. By using a systematic evolution of ligands by exponential enrichment (SELEX), different aptamers can be synthesized and each one targets one species of bacteria. By conjugating the aptamers onto a substrate, such as the microchannel in a microfluidic device, the mixture of bacterial cocktails can be individually captured by each aptamer that eventually achieve multiplex detection in a simultaneous manner. For example, *S.* Typhimurium and *S. aureus* were simultaneously identified using different aptamers in a sandwich-type tandem-SERS detection within 3 h [76]. Sandeep and co-workers proposed another simple and robust cross-platform approach using different nanoparticles functionalized with specific capturing ligands and Raman reporter molecules. This multiplex detection platform was applied for simultaneous detection of three different pathogens and offered an LOD ranging between $10^2$ and $10^3$ CFU/mL with a total detection time less than 45 min [64].

### 3.3. "Two-Step" and "One-Step" SERS

In comparison to the aforementioned concepts of "direct sensing" and "indirect sensing", "two-step sensing" and "one-step sensing" is another pair of the terminologies that are related to tandem-SERS platform. Once the separation and SERS detection are separate, it refers to "two-step" sensing. An intriguing "two-step" SERS approach based on a sandwich assay for the separation and detection of multiple pathogens in food samples was demonstrated by Wang and co-authors [70]. Figure 6a depicted the key steps of the process. The targeted pathogens in a food matrix were first captured and separated using silica-coated magnetic nanoparticles functionalized with the corresponding antibodies. Then, AuNPs integrated with a Raman reporter and surface-modified antibodies specific to the pathogen were used to complete the SERS detection. This platform achieved a LOD of $10^3$ CFU/mL for multiplex detection of *S.* Typhimurium and *S. aureus* in spinach wash water and peanut butter. "One step" sensing indicates that the separation and detection can occur simultaneously. Once "one-step" sensing is applied, a critical parameter is to ensure that the distance of the separation element is within 10 nm from the SERS-active substrate [90]; otherwise, the SERS effect will be tremendously reduced [101]. Naja and coauthors presented a "one-step" sensing of bacteria using silver nanoparticles functionalized with antibodies (Figure 6b). When the model bacteria attached to the corresponding antibodies absorbed on the protein-A-modified silver nanoparticles, the distance between the bacterium and the nanoparticle surface was 8 nm, thus the SERS signal of the bacterial cell wall would be generated and detected [102]. Further, "one-step" tandem-SERS sensing requires a relatively more complete clean-up of the sample matrices than that of the "two-step" tandem-SERS sensing method [90].

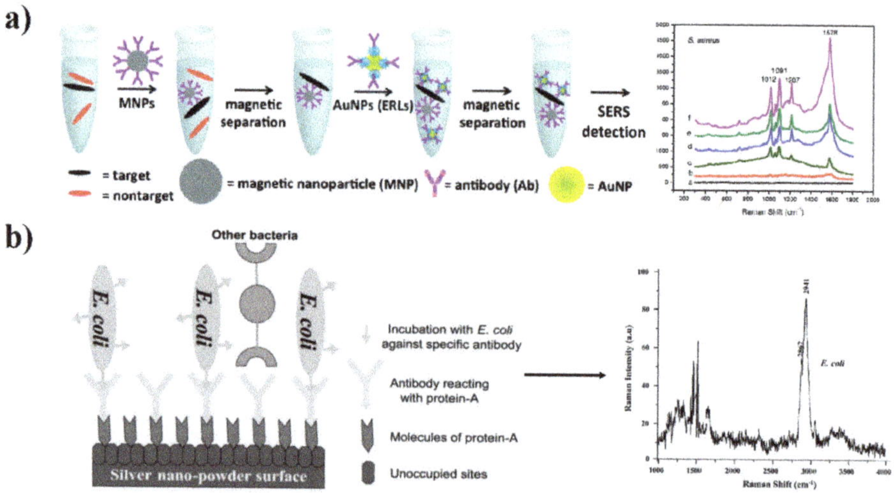


**Figure 6.** Representative "two-step" (**a**) and "one-step" (**b**) tandem-SERS sensing methods. (**a**) Tandem-SERS platform composed of the magnetic-based separation and SERS detection for multiple pathogens in food matrices. Pathogens were first captured with silica-coated magnetic probes, and then pathogen specific SERS probes (gold nanoparticles integrated with a Raman reporter and corresponding antibodies) were deployed to complete the following detection. (**b**) Schematic diagram for SERS-based detection of *E. coli* using silver nanoparticles conjugated with antibodies. Reproduced with permission [70]. Copyright Springer-Verlag, 2010. Reproduced with permission [102]. Copyright Royal Society of Chemistry, 2007.

## 4. Elucidating Antibiotic Resistant Mechanism of Bacteria Using SERS and Chemometrics

Besides the detection of antibiotic resistant bacteria either in a simple matrix or a complicated environmental, agri-food or clinical sample matrix, another major research direction of using SERS is to study the working mode and mechanism of antibiotics to inactivate bacteria. Bacterial cells can develop various strategies to resist to the antibiotic treatment as the pinnacle of evolution. Although new antibiotic resistance has been continuously emerging and spreading globally, bacteria use is one of two leading genetic strategies to deal with antibiotic treatment, namely mutation in genes associated with the action of antibiotic compounds and the acquisition of external DNA for the resistance determinants through horizontal gene transfer [103]. These genetic variations will lead to the change in biochemical composition of the bacterial cells. For example, three different biochemical routes can arise, fluoroquinolone resistance, including over-expression of efflux pumps to extrude the antibiotics from the bacterial cells, mutations in genes encoding DNA gyrase and topoisomerase, and generating specific proteins to protect the targeted site of fluoroquinolone [104,105].

### *4.1. Characterization of Antibiotic Resistance of Bacteria Using SERS*

As a three-dimensional complex surrounding the bacterial cells, peptidoglycan is the major component of the bacterial cell wall [106]. Since a relatively large amount of antibiotics is designed to target the bacterial cell wall, the biochemical compositions of the bacterial cell wall are expected to change along with the treatment of these antibiotics. Because SERS can record the macromolecular fingerprints of the bacterial cell membrane and cell wall, it can be applied to determine the effectiveness of antibiotic treatment as well as the antibiotic resistance patterns of the bacterial cells [15]. Although conventional Raman spectroscopy has been widely applied to profile the phenotypic response of bacteria to the antibiotic treatment, it requires a high concentration of bacterial culture for the collection of Raman signal [107]. Therefore, a relatively long time for bacterial cultivation and enrichment is necessary. By applying SERS for characterization, the antibiotic-resistant pattern of a single bacterial cell can be achieved. In addition, it will be critical to study the variations in responses among individual cells to the antibiotic treatment.

Antibiotic susceptibility testing (AST) is used to evaluate the effectiveness of antibiotic treatment against the pathogen infections. SERS-based AST could reduce the time by avoiding the need for overnight culture in MIC determination through the conventional AST methods. Liu and coauthors used an SERS-active substrate made of AgNPs imbedded in AAO to determine the antibiotic sensitivity of *E. coli* and *S. aureus* at the single-bacterium level [108]. Antibiotic-sensitive bacteria could be differentiated from antibiotic-resistant ones within 1 h after antibiotic exposure by monitoring the characteristic changes in SERS spectral profile. This approach demonstrated that SERS has the potential for direct detection and characterization of antibiotic resistance in real world samples instead of pure bacterial culture. Another study employed SERS-active AuNPs to study the antibiotic susceptibility of 12 urinary tract infection-causing bacteria [109]. Strain-specific identification was achieved with analytical sensitivity >95% and specificity >99%. The time for positive identification and AST was reduced to less than one hour.

In addition, SERS-active substrate can be employed as a means to establish MICs for various bacteria. Liu and colleagues demonstrated that SERS could monitor the reduction of specific bacterial biomarkers along with the treatment of antibiotics within two hours [110]. Clinical isolates of MRSA were exposed to vancomycin, while *E. coli*, *A. baumannii*, and *K. pneumoniae* were exposed to imipenem at the incremental concentrations. The isolates were determined as susceptible, intermediate, and resistant based on the change of the characteristic bands in SERS signals at a very early stage of antibiotic treatment, and the SERS MIC results were in excellent agreement with the standardized plate dilution methods that took upward of 24 h to complete. In a recent study, Cui and coauthors developed a homogeneous vacuum filtration-based method to improve SERS signal reproducibility and illustrated that the existence of heavy metal arsenic could increase the MIC of bacteria to the treatment of tetracycline. The authors claimed that SERS has the potential for culture-free characterization of resistome in a real microbiota system at the single cell sensitivity level [111].

Furthermore, monitoring the characteristic bacteria cell wall bands in the SERS spectra allowed for a further understanding of the antibiotic degradation mechanisms. The antibiotic response of *Lactococcus latis* was investigated using SERS-active AuNPs [112]. Antibiotic-induced spectral changes from ampicillin and ciprofloxacin were observed at 60 min after exposure to both antibiotics. However, ciprofloxacin induced only minor changes while ampicillin induced large SERS spectral changes. This was possibly because the inactivation mechanism of ciprofloxacin is to disrupt DNA synthesis, therefore the cell wall integrity was maintained for extended time periods and the cell wall signatures remained stable in the SERS spectra. While ampicillin interrupts the cell wall synthesis, which was directly detected by the SERS-active AuNPs. In another study, the SERS signals of *E. coli* were tracked upon antibiotic exposure to chloramphenicol, trimethoprim, polymyxin B, ampicillin, and formalin [113]. No spectral changes were observed after exposure to formalin although in vitro tests, which confirmed the cells were not viable. The authors noted that it was most likely due to the mechanism of formalin to crosslink membrane proteins but not degrade the cell wall. Similar results were observed with chloramphenicol and trimethoprim, which inactivate bacteria by inhibiting protein and DNA synthesis, respectively. The SERS signals remained unchanged after 2h exposure, which is possibly attributable to the sustained cell wall integrity. In contrast, SERS spectra changed within 5 min after antibiotic exposure to polymyxin B and ampicillin. They both aggressively degraded the bacterial cell wall, which released the SERS-active AgNPs and drastically reduced the SERS intensities. The technique could be used to further understand the fundamental mechanisms of microbial inactivation.

### 4.2. Chemometrics Used with SERS

Chemometric statistical analyses are usually required to decipher Raman spectral patterns so that minor variations in the spectral features of different biological samples can be distinguished. Multidimensional information of SERS spectra can be reduced into a few independent latent variables (called principal components) that account for the most variability of the original dataset by multivariate statistical analyses [114]. These principal components can then be used to segregate and quantify analytes based upon specific calibration models [115]. Chemometric methods include both unsupervised and supervised algorithms [116]. Among the spectroscopic-based pattern recognition methods, unsupervised principal component analysis (PCA) and hierarchical cluster analysis (HCA) are commonly used to provide either cluster plots or dendrogram structures for segregation and discrimination based upon the minor differences in Raman spectra [117]. Supervised chemometric models are generally used with some known answer from existing knowledge of the sample. Discriminant function analysis (DFA), partial least squares regression (PLSR), and soft independent modeling of class analog (SIMCA) are some of the most widely used models for the interpretation of SERS results [114]. For instance, a discriminant analysis is divided into two steps: to build a model using Raman spectra of bacterial cultures exposed to antibiotics of known class assignments, and to classify a new Raman spectrum of an antibiotic-exposed culture based on the distance to the multivariate mean of the closest class [118].

Different bacterial species or strains can be segregated into distinct groups based upon different biochemical compositions reflected by the major latent variables. For example, *E. coli*, *S. epidermidis* and four *Salmonella* strains exhibiting antibiotic resistance to the common therapeutics were detected and differentiated using SERS coupled with PCA [69]. In another work, SERS spectra of *P. mirabillis* and *Enterococcus* were quite similar despite having different cell wall structures. DFA was employed to analyze the subtle differences of SERS spectra from 6 strains of clinical urinary tract infection isolates for identification at genus-level [35]. Chemometric analysis play an important role in the determination of antibiotic resistance by SERS-based methods. Spectral differentiation of antibiotic resistant and sensitive strains can be demonstrated by chemometric models. For instance, Tien and others applied PCA for Raman spectra from MSSA and MRSA. MRSA cluster and MSSA cluster were segregated that can be used to differentiate MRSA from MSSA [119]. A SERS-based PLSR model was used to accurately determine the concentration of an MRSA strain in a mixture containing MSSA [54]. One recent study applied a three level chemometric model based on PLSR in combination with linear discriminant analysis (LDA) to extract those molecular changes and distinguish vancomycin-resistant and sensitive *Enterococci*. In addition, antibiotic-induced spectral changes from ampicillin and ciprofloxacin were monitored and statistically analyzed using PCA to understand the different working mechanisms of these antibiotics [112].

## 5. Conclusions and Future Direction

Raman spectroscopy and SERS have been validated for their potential in bacterial detection, typing, and characterization for almost three decades. Compared to the application of MALDI-TOF mass spectrometry for bacterial characterization, the use of Raman spectroscopy and SERS by industry is still in its infancy. This is mainly due to the relatively poor spectral reproducibility by using different types of the manufactured SERS substrates. As indicated in numerous review papers related to SERS bacterial study, to develop a stable SERS-active substrate for consistent and global use in a commercial manner is highly critical to promote this versatile technology to environmental, agri-food and clinical applications. Another major challenge is the relatively high cost of the confocal micro-Raman spectroscopic system. Although very little cost is required for purchasing consumables and instrumental maintenance compared to MALDI-TOF mass spectrometry, industries are still reluctant to purchase a bench-top Raman spectroscopic system. Therefore, a portable/handheld Raman instrument might be more affordable even though the resolution of the collected SERS spectra is relatively low. A more user-friendly software is also required for the convenient spectral interpretation as well as chemometric analyses. Several vendors have developed their own software for spectral processing and chemometrics, but a major doubt is how reliable such software for spectral analysis can be. By only clicking each "black-box" in the software, the researchers may not fully understand how each algorithm will affect the performance of the chemometric models. A standardized protocol for SERS spectral analyses and chemometric analyses therefore is critical to achieve inter-laboratory validation of the results for bacterial characterization, such as the characterization of bacterial antibiotic resistance.

Albeit these aforementioned challenges and potential limitations, SERS is definitely a very promising candidate for the determination of bacterial antibiotic resistance in a high-throughput, multiplex, and ultrafast manner. We suggest that industries use SERS for the detection and characterization of bacterial antibiotic resistance as an innovative fast screening alternative that can couple with the conventional methods for a further confirmation. Along with the further advancement in optical instrumentation and machine learning, the new version of the Raman spectroscopic system will be more user-friendly and cost-effective. We also envision that SERS can be used to further illustrate the modes of antibiotic and antimicrobial resistance of bacteria. This may contribute to the design of more effective antimicrobial treatment. Although SERS itself can be regarded as the core technology for an individual project, such as the detection of antibiotic resistance bacteria in a clinical specimen, we also believe it can be integrated as part of a more complicated study to drive very fundamental scientific research questions related to bacterial antibiotic resistance.

**Author Contributions:** K.W. is the leading author to draft this review paper along with additional contributions from other coauthors. K.W. and X.L. developed the structure of this review paper. S.L. and M.P. provided critical feedback and helped revise the paper. S.W. contributed to the final version of the manuscript.

**Funding:** This research was funded by Discovery Grant from the National Sciences and Engineering Research Council of Canada (NSERC RGPIN-2014-05487).

**Acknowledgments:** Financial support to X.L. in the form of a Discovery Grant from the National Sciences and Engineering Research Council of Canada (NSERC RGPIN-2014-05487) is gratefully acknowledged.

**Conflicts of Interest:** The authors declare no conflict of interest.

## References

1. Law, J.W.-F.; Ab Mutalib, N.-S.; Chan, K.-G.; Lee, L.-H. Rapid methods for the detection of foodborne bacterial pathogens: Principles, applications, advantages and limitations. *Front. Microbiol.* **2015**, *5*, 770. [CrossRef] [PubMed]
2. Papadakis, G.; Murasova, P.; Hamiot, A.; Tsougeni, K.; Kaprou, G.; Eck, M.; Rabus, D.; Bilkova, Z.; Dupuy, B.; Jobst, G. Micro-nano-bio acoustic system for the detection of foodborne pathogens in real samples. *Biosens. Bioelectron.* **2018**, *111*, 52–58. [CrossRef] [PubMed]
3. Velusamy, V.; Arshak, K.; Korostynska, O.; Oliwa, K.; Adley, C. An overview of foodborne pathogen detection: In the perspective of biosensors. *Biotechnol. Adv.* **2010**, *28*, 232–254. [CrossRef] [PubMed]
4. Dwivedi, H.P.; Jaykus, L.-A. Detection of pathogens in foods: The current state-of-the-art and future directions. *Crit. Rev. Microbiol.* **2011**, *37*, 40–63. [CrossRef] [PubMed]

5. Gracias, K.S.; McKillip, J.L. A review of conventional detection and enumeration methods for pathogenic bacteria in food. *Can. J. Microbiol.* **2004**, *50*, 883–890. [CrossRef] [PubMed]
6. Molina, F.; López-Acedo, E.; Tabla, R.; Roa, I.; Gómez, A.; Rebollo, J.E. Improved detection of escherichia coli and coliform bacteria by multiplex pcr. *BMC Biotechnol.* **2015**, *15*, 48. [CrossRef] [PubMed]
7. Carbonnelle, E.; Mesquita, C.; Bille, E.; Day, N.; Dauphin, B.; Beretti, J.-L.; Ferroni, A.; Gutmann, L.; Nassif, X. Maldi-tof mass spectrometry tools for bacterial identification in clinical microbiology laboratory. *Clin. Biochem.* **2011**, *44*, 104–109. [CrossRef] [PubMed]
8. Croxatto, A.; Prod'hom, G.; Greub, G. Applications of maldi-tof mass spectrometry in clinical diagnostic microbiology. *FEMS Microbiol. Rev.* **2012**, *36*, 380–407. [CrossRef] [PubMed]
9. Szabados, F.; Michels, M.; Kaase, M.; Gatermann, S. The sensitivity of direct identification from positive bact/alert™(biomérieux) blood culture bottles by matrix-assisted laser desorption ionization time-of-flight mass spectrometry is low. *Clin. Microbiol. Infect.* **2011**, *17*, 192–195. [CrossRef] [PubMed]
10. Blair, J.M.; Webber, M.A.; Baylay, A.J.; Ogbolu, D.O.; Piddock, L.J. Molecular mechanisms of antibiotic resistance. *Nat. Rev. Microbiol.* **2015**, *13*, 42. [CrossRef] [PubMed]
11. Bassetti, M.; Poulakou, G.; Ruppe, E.; Bouza, E.; Van Hal, S.J.; Brink, A. Antimicrobial resistance in the next 30 years, humankind, bugs and drugs: A visionary approach. *Intensive Care Med.* **2017**, *43*, 1464–1475. [CrossRef] [PubMed]
12. Strommenger, B.; Kettlitz, C.; Werner, G.; Witte, W. Multiplex pcr assay for simultaneous detection of nine clinically relevant antibiotic resistance genes in staphylococcus aureus. *J. Clin. Microbiol.* **2003**, *41*, 4089–4094. [CrossRef] [PubMed]
13. Gullberg, E.; Cao, S.; Berg, O.G.; Ilbäck, C.; Sandegren, L.; Hughes, D.; Andersson, D.I. Selection of resistant bacteria at very low antibiotic concentrations. *PLoS Pathog.* **2011**, *7*, e1002158. [CrossRef] [PubMed]
14. Pulido, M.R.; García-Quintanilla, M.; Martín-Peña, R.; Cisneros, J.M.; McConnell, M.J. Progress on the development of rapid methods for antimicrobial susceptibility testing. *J. Antimicrob. Chemother.* **2013**, *68*, 2710–2717. [CrossRef] [PubMed]
15. Galvan, D.D.; Yu, Q. Surface-enhanced raman scattering for rapid detection and characterization of antibiotic-resistant bacteria. *Adv. Healthc. Mater.* **2018**, 1701335. [CrossRef] [PubMed]
16. Wei, W.Y.; White, I.M. A simple filter-based approach to surface enhanced raman spectroscopy for trace chemical detection. *Analyst* **2012**, *137*, 1168–1173.
17. Kneipp, K.; Kneipp, H.; Bohr, H.G. Single-molecule SERS spectroscopy. In *Surface-Enhanced Raman Scattering*; Springer: Berlin, Germany, 2006; pp. 261–277.
18. Alvarez-Puebla, R.A.; Liz-Marzán, L.M. SERS detection of small inorganic molecules and ions. *Angew. Chem. Int. Ed.* **2012**, *51*, 11214–11223. [CrossRef] [PubMed]
19. An, Q.; Zhang, P.; Li, J.-M.; Ma, W.-F.; Guo, J.; Hu, J.; Wang, C.-C. Silver-coated magnetite–carbon core–shell microspheres as substrate-enhanced SERS probes for detection of trace persistent organic pollutants. *Nanoscale* **2012**, *4*, 5210–5216. [CrossRef] [PubMed]
20. Alvarez-Puebla, R.A.; Liz-Marzán, L.M. SERS-based diagnosis and biodetection. *Small* **2010**, *6*, 604–610. [CrossRef] [PubMed]
21. Kneipp, J.; Kneipp, H.; Kneipp, K. SERS—A single-molecule and nanoscale tool for bioanalytics. *Chem. Soc. Rev.* **2008**, *37*, 1052–1060. [CrossRef] [PubMed]
22. Kneipp, K.; Kneipp, H.; Manoharan, R.; Itzkan, I.; Dasari, R.R.; Feld, M.S. Surface-enhanced raman scattering (SERS)—A new tool for single molecule detection and identification. *Bioimaging* **1998**, *6*, 104–110. [CrossRef]
23. Chisanga, M.; Muhamadali, H.; Ellis, D.I.; Goodacre, R. Surface-enhanced raman scattering (SERS) in microbiology: Illumination and enhancement of the microbial world. *Appl. Spectrosc.* **2018**, *72*, 987–1000. [CrossRef] [PubMed]
24. McNay, G.; Eustace, D.; Smith, W.E.; Faulds, K.; Graham, D. Surface-enhanced raman scattering (SERS) and surface-enhanced resonance raman scattering (SERRS): A review of applications. *Appl. Spectrosc.* **2011**, *65*, 825–837. [CrossRef] [PubMed]
25. Le Ru, E.; Blackie, E.; Meyer, M.; Etchegoin, P.G. Surface enhanced raman scattering enhancement factors: A comprehensive study. *J. Phys. Chem. C* **2007**, *111*, 13794–13803. [CrossRef]
26. Fang, Y.; Seong, N.-H.; Dlott, D.D. Measurement of the distribution of site enhancements in surface-enhanced raman scattering. *Science* **2008**, *321*, 388–392. [CrossRef] [PubMed]

27. Valley, N.; Greeneltch, N.; Van Duyne, R.P.; Schatz, G.C. A look at the origin and magnitude of the chemical contribution to the enhancement mechanism of surface-enhanced raman spectroscopy (SERS): Theory and experiment. *J. Phys. Chem. Lett.* **2013**, *4*, 2599–2604. [CrossRef]
28. Stiles, P.L.; Dieringer, J.A.; Shah, N.C.; Van Duyne, R.P. Surface-enhanced raman spectroscopy. *Annu. Rev. Anal. Chem.* **2008**, *1*, 601–626. [CrossRef] [PubMed]
29. Willets, K.A.; Van Duyne, R.P. Localized surface plasmon resonance spectroscopy and sensing. *Annu. Rev. Phys. Chem.* **2007**, *58*, 267–297. [CrossRef] [PubMed]
30. Sharma, B.; Frontiera, R.R.; Henry, A.-I.; Ringe, E.; Van Duyne, R.P. SERS: Materials, applications, and the future. *Mater. Today* **2012**, *15*, 16–25. [CrossRef]
31. Radziuk, D.; Moehwald, H. Prospects for plasmonic hot spots in single molecule SERS towards the chemical imaging of live cells. *Phys. Chem. Chem. Phys.* **2015**, *17*, 21072–21093. [CrossRef] [PubMed]
32. Liu, Y.; Zhou, H.; Hu, Z.; Yu, G.; Yang, D.; Zhao, J. Label and label-free based surface-enhanced raman scattering for pathogen bacteria detection: A review. *Biosens. Bioelectron.* **2017**, *94*, 131–140. [CrossRef] [PubMed]
33. Jarvis, R.M.; Johnson, H.E.; Olembe, E.; Panneerselvam, A.; Malik, M.A.; Afzaal, M.; O'Brien, P.; Goodacre, R. Towards quantitatively reproducible substrates for SERS. *Analyst* **2008**, *133*, 1449–1452. [CrossRef] [PubMed]
34. Jarvis, R.M.; Goodacre, R. Characterisation and identification of bacteria using SERS. *Chem. Soc. Rev.* **2008**, *37*, 931–936. [CrossRef] [PubMed]
35. Jarvis, R.M.; Goodacre, R. Discrimination of bacteria using surface-enhanced raman spectroscopy. *Anal. Chem.* **2004**, *76*, 40–47. [CrossRef] [PubMed]
36. Jarvis, R.M.; Brooker, A.; Goodacre, R. Surface-enhanced raman scattering for the rapid discrimination of bacteria. *Faraday Discuss.* **2006**, *132*, 281–292. [CrossRef] [PubMed]
37. Xie, X.; Pu, H.; Sun, D.-W. Recent advances in nanofabrication techniques for SERS substrates and their applications in food safety analysis. *Crit. Rev. Food Sci. Nutr.* **2017**, *30*, 1–14. [CrossRef] [PubMed]
38. Gates, B.D.; Xu, Q.; Stewart, M.; Ryan, D.; Willson, C.G.; Whitesides, G.M. New approaches to nanofabrication: Molding, printing, and other techniques. *Chem. Rev.* **2005**, *105*, 1171–1196. [CrossRef] [PubMed]
39. Sharma, B.; Cardinal, M.F.; Kleinman, S.L.; Greeneltch, N.G.; Frontiera, R.R.; Blaber, M.G.; Schatz, G.C.; Van Duyne, R.P. High-performance SERS substrates: Advances and challenges. *MRS Bull.* **2013**, *38*, 615–624. [CrossRef]
40. Wang, P.; Pang, S.; Chen, J.; McLandsborough, L.; Nugen, S.R.; Fan, M.; He, L. Label-free mapping of single bacterial cells using surface-enhanced raman spectroscopy. *Analyst* **2016**, *141*, 1356–1362. [CrossRef] [PubMed]
41. Lu, Z.; Ruan, W.; Yang, J.; Xu, W.; Zhao, C.; Zhao, B. Deposition of ag nanoparticles on porous anodic alumina for surface enhanced raman scattering substrate. *J. Raman Spectrosc.* **2009**, *40*, 112–116. [CrossRef]
42. Ji, N.; Ruan, W.; Wang, C.; Lu, Z.; Zhao, B. Fabrication of silver decorated anodic aluminum oxide substrate and its optical properties on surface-enhanced raman scattering and thin film interference. *Langmuir* **2009**, *25*, 11869–11873. [CrossRef] [PubMed]
43. Ko, H.; Tsukruk, V.V. Nanoparticle-decorated nanocanals for surface-enhanced raman scattering. *Small* **2008**, *4*, 1980–1984. [CrossRef] [PubMed]
44. Pahlow, S.; Meisel, S.; Cialla-May, D.; Weber, K.; Rösch, P.; Popp, J. Isolation and identification of bacteria by means of raman spectroscopy. *Adv. Drug Deliv. Rev.* **2015**, *89*, 105–120. [CrossRef] [PubMed]
45. Kahraman, M.; Zamaleeva, A.I.; Fakhrullin, R.F.; Culha, M. Layer-by-layer coating of bacteria with noble metal nanoparticles for surface-enhanced raman scattering. *Anal. Bioanal. Chem.* **2009**, *395*, 2559. [CrossRef] [PubMed]
46. Zhou, H.; Yang, D.; Ivleva, N.P.; Mircescu, N.E.; Niessner, R.; Haisch, C. SERS detection of bacteria in water by in situ coating with ag nanoparticles. *Anal. Chem.* **2014**, *86*, 1525–1533. [CrossRef] [PubMed]
47. Chen, L.; Mungroo, N.; Daikuara, L.; Neethirajan, S. Label-free NIR-SERS discrimination and detection of foodborne bacteria by in situ synthesis of ag colloids. *J. Nanobiotechnol.* **2015**, *13*, 45. [CrossRef] [PubMed]
48. Mosier-Boss, P.; Sorensen, K.; George, R.; Obraztsova, A. SERS substrates fabricated using ceramic filters for the detection of bacteria. *Spectrochim. Acta Part A* **2016**, *153*, 591–598. [CrossRef] [PubMed]
49. Pu, H.; Xiao, W.; Sun, D.-W. SERS-microfluidic systems: A potential platform for rapid analysis of food contaminants. *Trends Food Sci. Technol.* **2017**, *70*, 114–126. [CrossRef]

50. Mungroo, N.A.; Oliveira, G.; Neethirajan, S. SERS based point-of-care detection of food-borne pathogens. *Microchim. Acta* **2016**, *183*, 697–707. [CrossRef]
51. Walter, A.; März, A.; Schumacher, W.; Rösch, P.; Popp, J. Towards a fast, high specific and reliable discrimination of bacteria on strain level by means of SERS in a microfluidic device. *Lab Chip* **2011**, *11*, 1013–1021. [CrossRef] [PubMed]
52. Xie, Y.; Yang, S.; Mao, Z.; Li, P.; Zhao, C.; Cohick, Z.; Huang, P.-H.; Huang, T.J. In situ fabrication of 3D Ag@ Zno nanostructures for microfluidic surface-enhanced raman scattering systems. *ACS Nano* **2014**, *8*, 12175–12184. [CrossRef] [PubMed]
53. Craig, A.P.; Franca, A.S.; Irudayaraj, J. Surface-enhanced raman spectroscopy applied to food safety. *Annu. Rev. Food Sci. Technol.* **2013**, *4*, 369–380. [CrossRef] [PubMed]
54. Lu, X.; Samuelson, D.R.; Xu, Y.; Zhang, H.; Wang, S.; Rasco, B.A.; Xu, J.; Konkel, M.E. Detecting and tracking nosocomial methicillin-resistant staphylococcus aureus using a microfluidic SERS biosensor. *Anal. Chem.* **2013**, *85*, 2320–2327. [CrossRef] [PubMed]
55. Mühlig, A.; Bocklitz, T.; Labugger, I.; Dees, S.; Henk, S.; Richter, E.; Andres, S.N.; Merker, M.; Stöckel, S.; Weber, K. Loc-SERS: A promising closed system for the identification of mycobacteria. *Anal. Chem.* **2016**, *88*, 7998–8004. [CrossRef] [PubMed]
56. Cheong, Y.; Kim, Y.J.; Kang, H.; Choi, S.; Lee, H.J. Rapid label-free identification of klebsiella pneumoniae antibiotic resistant strains by the drop-coating deposition surface-enhanced raman scattering method. *Spectrochim. Acta Part A* **2017**, *183*, 53–59. [CrossRef] [PubMed]
57. Zeiri, L.; Bronk, B.; Shabtai, Y.; Czege, J.; Efrima, S. Silver metal induced surface enhanced raman of bacteria. *Colloids Surf. A* **2002**, *208*, 357–362. [CrossRef]
58. Kahraman, M.; Yazıcı, M.M.; Şahin, F.; Çulha, M. Convective assembly of bacteria for surface-enhanced raman scattering. *Langmuir* **2008**, *24*, 894–901. [CrossRef] [PubMed]
59. Dina, N.; Zhou, H.; Colniţă, A.; Leopold, N.; Szoke-Nagy, T.; Coman, C.; Haisch, C. Rapid single-cell detection and identification of pathogens by using surface-enhanced raman spectroscopy. *Analyst* **2017**, *142*, 1782–1789. [CrossRef] [PubMed]
60. Wang, Y.; Lee, K.; Irudayaraj, J. Silver nanosphere SERS probes for sensitive identification of pathogens. *J. Phys. Chem. C* **2010**, *114*, 16122–16128. [CrossRef]
61. Wu, X.; Huang, Y.-W.; Park, B.; Tripp, R.A.; Zhao, Y. Differentiation and classification of bacteria using vancomycin functionalized silver nanorods array based surface-enhanced raman spectroscopy and chemometric analysis. *Talanta* **2015**, *139*, 96–103. [CrossRef] [PubMed]
62. Tamer, U.; Boyacı, İ.H.; Temur, E.; Zengin, A.; Dincer, I.; Elerman, Y. Fabrication of magnetic gold nanorod particles for immunomagnetic separation and SERS application. *J. Nanopart. Res.* **2011**, *13*, 3167–3176. [CrossRef]
63. Xiao, N.; Wang, C.; Yu, C. A self-referencing detection of microorganisms using surface enhanced raman scattering nanoprobes in a test-in-a-tube platform. *Biosensors* **2013**, *3*, 312–326. [CrossRef] [PubMed]
64. Ravindranath, S.P.; Wang, Y.; Irudayaraj, J. SERS driven cross-platform based multiplex pathogen detection. *Sens. Actuators B* **2011**, *152*, 183–190. [CrossRef]
65. Ondera, T.J.; Hamme, A.T., II. A gold nanopopcorn attached single-walled carbon nanotube hybrid for rapid detection and killing of bacteria. *J. Mater. Chem. B* **2014**, *2*, 7534–7543. [CrossRef] [PubMed]
66. Jones, S.; Sinha, S.S.; Pramanik, A.; Ray, P.C. Three-dimensional (3D) plasmonic hot spots for label-free sensing and effective photothermal killing of multiple drug resistant superbugs. *Nanoscale* **2016**, *8*, 18301–18308. [CrossRef] [PubMed]
67. Madiyar, F.R.; Bhana, S.; Swisher, L.Z.; Culbertson, C.T.; Huang, X.; Li, J. Integration of a nanostructured dielectrophoretic device and a surface-enhanced raman probe for highly sensitive rapid bacteria detection. *Nanoscale* **2015**, *7*, 3726–3736. [CrossRef] [PubMed]
68. Yakes, B.J.; Lipert, R.J.; Bannantine, J.P.; Porter, M.D. Detection of mycobacterium avium subsp. Paratuberculosis by a sonicate immunoassay based on surface-enhanced raman scattering. *Clin. Vaccine Immunol.* **2008**, *15*, 227–234. [CrossRef] [PubMed]
69. Wu, X.; Xu, C.; Tripp, R.A.; Huang, Y.-W.; Zhao, Y. Detection and differentiation of foodborne pathogenic bacteria in mung bean sprouts using field deployable label-free SERS devices. *Analyst* **2013**, *138*, 3005–3012. [CrossRef] [PubMed]

70. Wang, Y.; Ravindranath, S.; Irudayaraj, J. Separation and detection of multiple pathogens in a food matrix by magnetic SERS nanoprobes. *Anal. Bioanal. Chem.* **2011**, *399*, 1271–1278. [CrossRef] [PubMed]
71. Liu, X.; Knauer, M.; Ivleva, N.P.; Niessner, R.; Haisch, C. Synthesis of core-shell surface-enhanced raman tags for bioimaging. *Anal. Chem.* **2009**, *82*, 441–446. [CrossRef] [PubMed]
72. Khan, S.A.; Singh, A.K.; Senapati, D.; Fan, Z.; Ray, P.C. Targeted highly sensitive detection of multi-drug resistant salmonella DT104 using gold nanoparticles. *Chem. Commun.* **2011**, *47*, 9444–9446. [CrossRef] [PubMed]
73. Duan, N.; Chang, B.; Zhang, H.; Wang, Z.; Wu, S. Salmonella typhimurium detection using a surface-enhanced raman scattering-based aptasensor. *Int. J. Food Microbiol.* **2016**, *218*, 38–43. [CrossRef] [PubMed]
74. Duan, N.; Yan, Y.; Wu, S.; Wang, Z. Vibrio parahaemolyticus detection aptasensor using surface-enhanced raman scattering. *Food Control* **2016**, *63*, 122–127. [CrossRef]
75. Ma, S.; Cai, Q.; Lu, K.; Liao, F.; Shao, M. Bi-functional au/fes (Au/$Co_3O_4$) composite for in situ SERS monitoring and degradation of organic pollutants. *J. Nanopart. Res.* **2016**, *18*, 26. [CrossRef]
76. Zhang, H.; Ma, X.; Liu, Y.; Duan, N.; Wu, S.; Wang, Z.; Xu, B. Gold nanoparticles enhanced SERS aptasensor for the simultaneous detection of salmonella typhimurium and staphylococcus aureus. *Biosens. Bioelectron.* **2015**, *74*, 872–877. [CrossRef] [PubMed]
77. Liu, Z.; Wang, Y.; Deng, R.; Yang, L.; Yu, S.; Xu, S.; Xu, W. $Fe_3O_4$@ graphene oxide@ Ag particles for surface magnet solid-phase extraction surface-enhanced raman scattering (smspe-SERS): From sample pretreatment to detection all-in-one. *ACS Appl. Mater. Interfaces* **2016**, *8*, 14160–14168. [CrossRef] [PubMed]
78. Huang, P.J.; Tay, L.L.; Tanha, J.; Ryan, S.; Chau, L.K. Single-domain antibody-conjugated nanoaggregate-embedded beads for targeted detection of pathogenic bacteria. *Chem. Eur. J.* **2009**, *15*, 9330–9334. [CrossRef] [PubMed]
79. Catala, C.; Mir-Simon, B.; Feng, X.; Cardozo, C.; Pazos-Perez, N.; Pazos, E.; Gómez-de Pedro, S.; Guerrini, L.; Soriano, A.; Vila, J. Online SERS quantification of staphylococcus aureus and the application to diagnostics in human fluids. *Adv. Mater. Technol.* **2016**, *1*, 1600163. [CrossRef]
80. Drake, P.; Jiang, P.-S.; Chang, H.-W.; Su, S.-C.; Tanha, J.; Tay, L.-L.; Chen, P.; Lin, Y.-J. Raman based detection of staphylococcus aureus utilizing single domain antibody coated nanoparticle labels and magnetic trapping. *Anal. Methods* **2013**, *5*, 4152–4158. [CrossRef]
81. Qiu, L.; Wang, W.; Zhang, A.; Zhang, N.; Lemma, T.; Ge, H.; Toppari, J.J.; Hytönen, V.P.; Wang, J. Core–shell nanorod columnar array combined with gold nanoplate–nanosphere assemblies enable powerful in situ SERS detection of bacteria. *ACS Appl. Mater. Interfaces* **2016**, *8*, 24394–24403. [CrossRef] [PubMed]
82. Li, J.; Wang, C.; Kang, H.; Shao, L.; Hu, L.; Xiao, R.; Wang, S.; Gu, B. Label-free identification carbapenem-resistant escherichia coli based on surface-enhanced resonance raman scattering. *RSC Adv.* **2018**, *8*, 4761–4765. [CrossRef]
83. Liu, T.-Y.; Tsai, K.-T.; Wang, H.-H.; Chen, Y.; Chen, Y.-H.; Chao, Y.-C.; Chang, H.-H.; Lin, C.-H.; Wang, J.-K.; Wang, Y.-L. Functionalized arrays of raman-enhancing nanoparticles for capture and culture-free analysis of bacteria in human blood. *Nat. Commun.* **2011**, *2*, 538. [CrossRef] [PubMed]
84. Gu, H.; Xu, K.; Xu, C.; Xu, B. Biofunctional magnetic nanoparticles for protein separation and pathogen detection. *Chem. Commun.* **2006**, 941–949. [CrossRef] [PubMed]
85. Zhang, L.; Xu, J.; Mi, L.; Gong, H.; Jiang, S.; Yu, Q. Multifunctional magnetic–plasmonic nanoparticles for fast concentration and sensitive detection of bacteria using SERS. *Biosens. Bioelectron.* **2012**, *31*, 130–136. [CrossRef] [PubMed]
86. Wang, C.; Wang, J.; Li, M.; Qu, X.; Zhang, K.; Rong, Z.; Xiao, R.; Wang, S. A rapid SERS method for label-free bacteria detection using polyethylenimine-modified au-coated magnetic microspheres and Au@ Ag nanoparticles. *Analyst* **2016**, *141*, 6226–6238. [CrossRef] [PubMed]
87. Wang, J.; Wu, X.; Wang, C.; Rong, Z.; Ding, H.; Li, H.; Li, S.; Shao, N.; Dong, P.; Xiao, R. Facile synthesis of au-coated magnetic nanoparticles and their application in bacteria detection via a SERS method. *ACS Appl. Mater. Interfaces* **2016**, *8*, 19958–19967. [CrossRef] [PubMed]
88. Kearns, H.; Goodacre, R.; Jamieson, L.E.; Graham, D.; Faulds, K. SERS detection of multiple antimicrobial-resistant pathogens using nanosensors. *Anal. Chem.* **2017**, *89*, 12666–12673. [CrossRef] [PubMed]
89. Temur, E.; Boyacı, İ.H.; Tamer, U.; Unsal, H.; Aydogan, N. A highly sensitive detection platform based on surface-enhanced raman scattering for escherichia coli enumeration. *Anal. Bioanal. Chem.* **2010**, *397*, 1595–1604. [CrossRef] [PubMed]

90. Liao, W.; Lu, X. Determination of chemical hazards in foods using surface-enhanced raman spectroscopy coupled with advanced separation techniques. *Trends Food Sci. Technol.* **2016**, *54*, 103–113. [CrossRef]
91. Lin, Y.; Hamme, A.T., II. Targeted highly sensitive detection/eradication of multi-drug resistant salmonella dt104 through gold nanoparticle–swcnt bioconjugated nanohybrids. *Analyst* **2014**, *139*, 3702–3705. [CrossRef] [PubMed]
92. Yang, L. A review of multifunctions of dielectrophoresis in biosensors and biochips for bacteria detection. *Anal. Lett.* **2012**, *45*, 187–201. [CrossRef]
93. Lin, H.Y.; Huang, C.H.; Hsieh, W.H.; Liu, L.H.; Lin, Y.C.; Chu, C.C.; Wang, S.T.; Kuo, I.T.; Chau, L.K.; Yang, C.Y. On-line SERS detection of single bacterium using novel SERS nanoprobes and a microfluidic dielectrophoresis device. *Small* **2014**, *10*, 4700–4710. [CrossRef] [PubMed]
94. Ranjith Premasiri, W.; Lemler, P.; Chen, Y.; Gebregziabher, Y.; Ziegler, L.D. SERS analysis of bacteria, human blood, and cancer cells: A metabolomic and diagnostic tool. *Front. Surf.-Enhanc. Raman Scatt.* **2014**, 257–283. [CrossRef]
95. Zhang, J.Y.; Do, J.; Premasiri, W.R.; Ziegler, L.D.; Klapperich, C.M. Rapid point-of-care concentration of bacteria in a disposable microfluidic device using meniscus dragging effect. *Lab Chip* **2010**, *10*, 3265–3270. [CrossRef] [PubMed]
96. Hou, D.; Maheshwari, S.; Chang, H.-C. Rapid bioparticle concentration and detection by combining a discharge driven vortex with surface enhanced raman scattering. *Biomicrofluidics* **2007**, *1*, 014106. [CrossRef] [PubMed]
97. Guven, B.; Basaran-Akgul, N.; Temur, E.; Tamer, U.; Boyacı, İ.H. Sers-based sandwich immunoassay using antibody coated magnetic nanoparticles for escherichia coli enumeration. *Analyst* **2011**, *136*, 740–748. [CrossRef] [PubMed]
98. Lin, C.-C.; Yang, Y.-M.; Liao, P.-H.; Chen, D.-W.; Lin, H.-P.; Chang, H.-C. A filter-like aunps@ ms SERS substrate for staphylococcus aureus detection. *Biosens. Bioelectron.* **2014**, *53*, 519–527. [CrossRef] [PubMed]
99. Kamińska, A.; Witkowska, E.; Kowalska, A.; Skoczyńska, A.; Gawryszewska, I.; Guziewicz, E.; Snigurenko, D.; Waluk, J. Highly efficient SERS-based detection of cerebrospinal fluid neopterin as a diagnostic marker of bacterial infection. *Anal. Bioanal. Chem.* **2016**, *408*, 4319–4327. [CrossRef] [PubMed]
100. Witkowska, E.; Szymborski, T.; Kamińska, A.; Waluk, J. Polymer mat prepared via forcespinning™ as a SERS platform for immobilization and detection of bacteria from blood plasma. *Mater. Sci. Eng. C* **2017**, *71*, 345–350. [CrossRef] [PubMed]
101. Granger, J.H.; Schlotter, N.E.; Crawford, A.C.; Porter, M.D. Prospects for point-of-care pathogen diagnostics using surface-enhanced raman scattering (SERS). *Chem. Soc. Rev.* **2016**, *45*, 3865–3882. [CrossRef] [PubMed]
102. Naja, G.; Bouvrette, P.; Hrapovic, S.; Luong, J.H. Raman-based detection of bacteria using silver nanoparticles conjugated with antibodies. *Analyst* **2007**, *132*, 679–686. [CrossRef] [PubMed]
103. Levy, S.B.; Marshall, B. Antibacterial resistance worldwide: Causes, challenges and responses. *Nat. Med.* **2004**, *10*, S122. [CrossRef] [PubMed]
104. Wang, H.; Dzink-Fox, J.L.; Chen, M.; Levy, S.B. Genetic characterization of highly fluoroquinolone-resistant clinical escherichia coli strains from china: Role ofacrr mutations. *Antimicrob. Agents Chemother.* **2001**, *45*, 1515–1521. [CrossRef] [PubMed]
105. Schneiders, T.; Amyes, S.; Levy, S. Role of acrr and rama in fluoroquinolone resistance in clinical klebsiella pneumoniae isolates from singapore. *Antimicrob. Agents Chemother.* **2003**, *47*, 2831–2837. [CrossRef] [PubMed]
106. Liechti, G.; Kuru, E.; Hall, E.; Kalinda, A.; Brun, Y.; VanNieuwenhze, M.; Maurelli, A. A new metabolic cell-wall labelling method reveals peptidoglycan in chlamydia trachomatis. *Nature* **2014**, *506*, 507. [CrossRef] [PubMed]
107. Zu, T.N.; Athamneh, A.I.; Senger, R.S. A study of the phenotypic responses of escherichia coli to multiple 4-carbon alcohols using raman spectroscopy. In *Phenotypic And Metabolic Profiling Of Biological Samples In Near Real-Time Using Raman Spectroscopy*; Virginia Tech: Blacksburg, VA, USA, 2014; Volume 1001, p. 78.
108. Liu, T.-T.; Lin, Y.-H.; Hung, C.-S.; Liu, T.-J.; Chen, Y.; Huang, Y.-C.; Tsai, T.-H.; Wang, H.-H.; Wang, D.-W.; Wang, J.-K. A high speed detection platform based on surface-enhanced raman scattering for monitoring antibiotic-induced chemical changes in bacteria cell wall. *PLoS ONE* **2009**, *4*, e5470. [CrossRef] [PubMed]
109. Premasiri, W.; Chen, Y.; Williamson, P.; Bandarage, D.; Pyles, C.; Ziegler, L. Rapid urinary tract infection diagnostics by surface-enhanced raman spectroscopy (SERS): Identification and antibiotic susceptibilities. *Anal. Bioanal. Chem.* **2017**, *409*, 3043–3054. [CrossRef] [PubMed]

110. Liu, C.-Y.; Han, Y.-Y.; Shih, P.-H.; Lian, W.-N.; Wang, H.-H.; Lin, C.-H.; Hsueh, P.-R.; Wang, J.-K.; Wang, Y.-L. Rapid bacterial antibiotic susceptibility test based on simple surface-enhanced raman spectroscopic biomarkers. *Sci. Rep.* **2016**, *6*, 23375. [CrossRef] [PubMed]
111. Cui, L.; Zhang, Y.-J.; Huang, W.E.; Zhang, B.-F.; Martin, F.L.; Li, J.-Y.; Zhang, K.-S.; Zhu, Y.-G. Surface-enhanced raman spectroscopy for identification of heavy metal arsenic (v)-mediated enhancing effect on antibiotic resistance. *Anal. Chem.* **2016**, *88*, 3164–3170. [CrossRef] [PubMed]
112. Wang, P.; Pang, S.; Zhang, H.; Fan, M.; He, L. Characterization of lactococcus lactis response to ampicillin and ciprofloxacin using surface-enhanced raman spectroscopy. *Anal. Bioanal. Chem.* **2016**, *408*, 933–941. [CrossRef] [PubMed]
113. Zhou, H.; Yang, D.; Ivleva, N.P.; Mircescu, N.E.; Schubert, S.R.; Niessner, R.; Wieser, A.; Haisch, C. Label-free in situ discrimination of live and dead bacteria by surface-enhanced raman scattering. *Anal. Chem.* **2015**, *87*, 6553–6561. [CrossRef] [PubMed]
114. Lu, X.; Al-Qadiri, H.M.; Lin, M.; Rasco, B.A. Application of mid-infrared and raman spectroscopy to the study of bacteria. *Food Bioprocess Technol.* **2011**, *4*, 919–935. [CrossRef]
115. Ellis, D.I.; Goodacre, R. Metabolic fingerprinting in disease diagnosis: Biomedical applications of infrared and raman spectroscopy. *Analyst* **2006**, *131*, 875–885. [CrossRef] [PubMed]
116. Goodacre, R. Explanatory analysis of spectroscopic data using machine learning of simple, interpretable rules. *Vib. Spectrosc.* **2003**, *32*, 33–45. [CrossRef]
117. Lu, X.; Huang, Q.; Miller, W.G.; Aston, D.E.; Xu, J.; Xue, F.; Zhang, H.; Rasco, B.A.; Wang, S.; Konkel, M.E. Comprehensive detection and discrimination of campylobacter species by use of confocal micro-raman spectroscopy and multilocus sequence typing. *J. Clin. Microbiol.* **2012**, *50*, 2932–2946. [CrossRef] [PubMed]
118. Athamneh, A.; Alajlouni, R.; Wallace, R.; Seleem, M.; Senger, R. Phenotypic profiling of antibiotic response signatures in escherichia coli using raman spectroscopy. *Antimicrob. Agents Chemother.* **2014**, *58*, 1302–1314. [CrossRef] [PubMed]
119. Tien, N.; Chen, H.-C.; Gau, S.-L.; Lin, T.-H.; Lin, H.-S.; You, B.-J.; Tsai, P.-C.; Chen, I.-R.; Tsai, M.-F.; Wang, I.-K. Diagnosis of bacterial pathogens in the dialysate of peritoneal dialysis patients with peritonitis using surface-enhanced raman spectroscopy. *Clin. Chim. Acta* **2016**, *461*, 69–75. [CrossRef] [PubMed]

MDPI
St. Alban-Anlage 66
4052 Basel
Switzerland
Tel. +41 61 683 77 34
Fax +41 61 302 89 18
www.mdpi.com

*Nanomaterials* Editorial Office
E-mail: nanomaterials@mdpi.com
www.mdpi.com/journal/nanomaterials